GL BAL WARMING

Climate Management

Iroquois Ridge
High School
Resource Centre

Climate Management

Solving the Problem

Julie Kerr Casper, Ph.D.

CLIMATE MANAGEMENT: Solving the Problem

Copyright © 2010 by Julie Kerr Casper, Ph.D.

All rights reserved. No part of this book may be reproduced or utilized in any form or by any means, electronic or mechanical, including photocopying, recording, or by any information storage or retrieval systems, without permission in writing from the publisher. For information contact:

Facts On File, Inc.
An imprint of Infobase Publishing
132 West 31st Street
New York NY 10001

Library of Congress Cataloging-in-Publication Data

Casper, Julie Kerr.
   Climate management : solving the problem / Julie Kerr Casper.
      p. cm.—(Global warming)
   Includes bibliographical references and index.
   ISBN 978-0-8160-7266-8 (alk. paper)
1. Climatology—Popular works. 2. Global warming—Popular works. I. Title.

QC863.4.C36 2009
363.738'74—dc22   2009021254

Facts On File books are available at special discounts when purchased in bulk quantities for businesses, associations, institutions, or sales promotions. Please call our Special Sales Department in New York at (212) 967-8800 or (800) 322-8755.

You can find Facts On File on the World Wide Web at http://www.factsonfile.com

Text design by Erik Lindstrom
Illustrations by Dale Williams
Photo research by the author
Composition by Hermitage Publishing Services
Cover printed by Bang Printing, Brainerd, MN
Book printed and bound by Bang Printing, Brainerd, MN
Date printed: April 2010
Printed in the United States of America

10 9 8 7 6 5 4 3 2 1

This book is printed on acid-free paper and contains 30 percent postconsumer recycled content.

CONTENTS

Preface	viii
Acknowledgments	xii
Introduction	xiv

1	**The Beginning of Global Warming Management**	**1**
	The Human Link	2
	The United Nations Framework Convention on Climate Change and the Kyoto Protocol	4
	The Impacts of Warming in the United States and Canada	5
	The U.S. Response and International Reactions	14
	The Intergovernmental Panel on Climate Change	20
	IPCC Reports	25

2	**The U.S. Political Arena**	**32**
	The Current Political Climate	32
	President Obama and His Outlook on Global Warming	36
	National Security and Terrorism	40
	Current Legislation	44

3	**Cap and Trade and Other Mitigation Strategies**	**55**
	Cap and Trade	55
	State Mitigation Projects	62
	Economics of Mitigation	68

| 4 | **The International Political Arena** | **72** |
| | An Expert's Warning | 72 |

The Evolution of International Cooperation	75
The Role of International Organizations	78
The Progress of Individual Countries	84

5 Global Warming, Human Psychology, and the Media — 92

Human Psychology and Cultural Values	92
The Power of the Media	95
Keeping a Journalistic Balance	97
Scientists' Mindsets and Data Change	100

6 The Stand on the Debate — 103

Understanding Modern Climate	104
The Far Right—Skeptics of Global Warming	109
The Middle Ground	115
The Far Left—Believers in Global Warming	116

7 Green Energy and Global Warming Research — 122

The Environmental Benefits of Green Energy	122
Solar Energy	130
New Ways to Store Solar Energy	*137*
Geothermal Energy	139
Wind Energy	142
Hydropower	146
Energy from Biomass	148
Biofuel Crop Bans in Europe	*151*
Ocean Energy	152
Global Warming Research	155

8 Climate Modeling — 159

The Modeling Challenge—A Brief History	160
Fundamentals of Climate Modeling	162
Watching Earth's Climate Change in the Classroom	*171*
Modeling Uncertainties and Challenges	172

9 Practical Solutions That Work—Getting Everyone Involved — 182

Taking Action	183
Practical Solutions to Global Warming	184

The 2007 Nobel Peace Prize	*185*
Suggested Solutions That Are Not So Practical	193
Prioritizing Adaptation Strategies	198
Simple Activities Everyone Can Do	199

10 The Future: What Lies Ahead — 210

A Look toward the Future	211
Winners and Losers	216
New Technologies	219
The Final Choice	222

Appendixes	**223**
Chronology	**233**
Glossary	**239**
Further Resources	**247**
Index	**255**

PREFACE

*We do not inherit the Earth from our ancestors—
we borrow it from our children.*

This ancient Native American proverb and what it implies resonates today as it has become increasingly obvious that people's actions and interactions with the environment affect not only living conditions now, but also those of many generations to follow. Humans must address the effect they have on the Earth's climate and how their choices today will have an impact on future generations.

Many years ago, Mark Twain joked that "Everyone talks about the weather, but no one does anything about it." That is not true anymore. Humans are changing the world's climate and with it the local, regional, and global weather. Scientists tell us that "climate is what we expect, and weather is what we get." Climate change occurs when that average weather shifts over the long term in a specific location, a region, or the entire planet.

Global warming and climate change are urgent topics. They are discussed on the news, in conversations, and are even the subjects of horror movies. How much is fact? What does global warming mean to individuals? What should it mean?

The readers of this multivolume set—most of whom are today's middle and high school students—will be tomorrow's leaders and scientists. Global warming and its threats are real. As scientists unlock the mysteries of the past and analyze today's activities, they warn that future generations may be in jeopardy. There is now overwhelming evidence that human activities are changing the world's climate. For thousands of years, the Earth's atmosphere has changed very little; but today, there are problems in keeping the balance. Greenhouse gases are being added to the atmosphere at an alarming rate. Since the Industrial Revolution (late 18th, early 19th centuries), human activities from transportation, agriculture, fossil fuels, waste disposal and treatment, deforestation, power stations, land use, biomass burning, and industrial processes, among other things, have added to the concentrations of greenhouse gases.

These activities are changing the atmosphere more rapidly than humans have ever experienced before. Some people think that warming the Earth's atmosphere by a few degrees is harmless and could have no effect on them; but global warming is more than just a warming—or cooling—trend. Global warming could have far-reaching and unpredictable environmental, social, and economic consequences. The following demonstrates what a few degrees' change in the temperature can do.

The Earth experienced an ice age 13,000 years ago. Global temperatures then warmed up 8.3°F (5°C) and melted the vast ice sheets that covered much of the North American continent. Scientists today predict that average temperatures could rise 11.7°F (7°C) during this century alone. What will happen to the remaining glaciers and ice caps?

If the temperatures rise as leading scientists have predicted, less freshwater will be available—and already one-third of the world's population (about 2 billion people) suffer from a shortage of water. Lack of water will keep farmers from growing food. It will also permanently destroy sensitive fish and wildlife habitat. As the ocean levels rise, coastal lands and islands will be flooded and destroyed. Heat waves could kill tens of thousands of people. With warmer temperatures, outbreaks of diseases will spread and intensify. Plant pollen mold spores in the air will increase, affecting those with allergies. An increase in severe weather could result in hurricanes similar or even stronger than Katrina in 2005, which destroyed large areas of the southeastern United States.

Higher temperatures will cause other areas to dry out and become tinder for larger and more devastating wildfires that threaten forests, wildlife, and homes. If drought destroys the rain forests, the Earth's delicate oxygen and carbon balances will be harmed, affecting the water, air, vegetation, and all life.

Although the United States has been one of the largest contributors to global warming, it ranks far below countries and regions—such as Canada, Australia, and western Europe—in taking steps to fix the damage that has been done. Global Warming is a multivolume set that explores the concept that each person is a member of a global family who shares responsibility for fixing this problem. In fact, the only way to fix it is to work together toward a common goal. This seven-volume set covers all of the important climatic issues that need to be addressed in order to understand the problem, allowing the reader to build a solid foundation of knowledge and to use the information to help solve the critical issues in effective ways. The set includes the following volumes:

Climate Systems
Global Warming Trends
Global Warming Cycles
Changing Ecosystems
Greenhouse Gases
Fossil Fuels and Pollution
Climate Management

These volumes explore a multitude of topics—how climates change, learning from past ice ages, natural factors that trigger global warming on Earth, whether the Earth can expect another ice age in the future, how the Earth's climate is changing now, emergency preparedness in severe weather, projections for the future, and why climate affects everything people do from growing food, to heating homes, to using the Earth's natural resources, to new scientific discoveries. They look at the impact that rising sea levels will have on islands and other areas worldwide, how individual ecosystems will be affected, what humans will lose if rain forests are destroyed, how industrialization and pollution puts peoples' lives at risk, and the benefits of developing environmentally friendly energy resources.

The set also examines the exciting technology of computer modeling and how it has unlocked mysteries about past climate change and global warming and how it can predict the local, regional, and global climates of the future—the very things leaders of tomorrow need to know *today*.

We will know only what we are taught;
We will be taught only what others deem is important to know;
And we will learn to value that which is important.
—Native American proverb

ACKNOWLEDGMENTS

Global warming may very well be one of the most important issues you will have to make a decision on in your lifetime. The decisions you make on energy sources and daily conservation practices will determine not only the quality of your life, but also the lives of your descendants.

I cannot stress enough how important it is to gain a good understanding of global warming: what it is, why it is happening, how it can be slowed down, why everybody is contributing to the problem, and why *everybody* needs to be an active part of the solution.

I would sincerely like to thank several of the federal government agencies that research, educate, and actively take part in dealing with the global warming issue—in particular, the National Aeronautics and Space Administration (NASA), the National Oceanic and Atmospheric Administration (NOAA), the Environmental Protection Agency (EPA), and the U.S. Geological Survey (USGS) for providing an abundance of resources and outreach programs on this important subject. I give special thanks to James E. Hansen, Al Gore, and Arnold Schwarzenegger for their diligent efforts toward bringing the global warming issue so powerfully to the public's attention. I would also like to acknowledge and give thanks to the many wonderful universities across the United States, in England, Canada, and Australia, as well as private organizations, such as the World Wildlife Fund, that diligently strive to educate others and help toward finding a solution to this very real problem.

Acknowledgments

I want to give a huge thanks to my agent, Jodie Rhodes, for her assistance, guidance, and efforts; Frank K. Darmstadt, executive editor, for all his hard work, dedication, support, and helpful advice and attention to detail; Alexandra Lo Re, assistant editor; and the production department for their assistance and the outstanding quality of their work.

INTRODUCTION

Whether people argue that global warming is caused by natural phenomena or by humans—or both—it is one of the most controversial topics in the scientific world today. Scientists have many theories about global warming and because there are so many factors involved, it is difficult to pinpoint exactly what causes it and what is to be done. Earth's climate is extremely complicated, and *climatologists* are conducting daily research in order to improve their understanding of all the interrelated components.

Each year, about 7 billion tons (6.4 metric tons) of carbon is released into the atmosphere. Studies show that *concentrations* of *carbon dioxide* have increased by about one-third since 1900. During this same time period, experts say the Earth has warmed rapidly. Many scientists believe this means that humans are contributing significantly to global warming. Even scientists who are skeptical about global warming recognize that there is much more carbon dioxide in the atmosphere than ever before.

Many natural events can cause the Earth's climate to change, such as shifting ocean currents, changes in the amount of solar energy that reaches the Earth, and the eruption of volcanoes. Human-induced causes are also critical, such as burning fossil fuels and polluting the atmosphere. Today, there is a wealth of evidence supporting global warming: measured increases in average temperatures; changing rainfall patterns; rising sea levels; glaciers thinning and retreating; coral reefs dying as oceans become warmer; more frequent droughts in Africa and Asia; *permafrost* melting in the Arctic; lakes and rivers that freeze during the winter thawing earlier; plants and animals shifting habitat ranges toward the polar regions and to higher altitudes on mountains; and disrupted migration patterns for wildlife, such as polar bears, whales, and the monarch butterfly.

One of scientists' biggest concerns about global warming is that the real danger is unknown. Because climate is such a complicated system, there are still a lot of areas in climatology that are not well understood. As computer *models* become more sophisticated and instruments are developed that can identify and monitor specific portions of the atmosphere, experts will have a better idea on how best to manage the human impact on climate. One thing is certain—human behavior has an enormous impact on global warming.

What is desperately needed at this point is for countries to begin working together to solve these issues. Not just a handful of countries can fix climate change—it is a global problem. Developed countries such as the United States or Great Britain are more financially able to make changes, however, a way must be found to bring all nations of the world into the discussion and solution of climate change. *Climate Management* focuses on changing human behavior as the first step toward fixing the problem to keep the world from suffering the disastrous effects of climate change. The book discusses the role of the United Nations in the effort to manage global warming and explores human psychology and how cultural values, politics, and news dissemination can affect people's opinions, thereby driving public response. The book also informs the reader how global warming affects national security and why its progression is a very real threat to everyone's future.

Climate Management outlines the contributions and successes of the *IPCC* and the Kyoto *Protocol* to the worldwide effort to control global warming and explains why a global push to combat this problem is the only way any solution will be found. The book also presents various conservation programs that have been developed by individual countries, international committees, private organizations, and individuals.

It looks at what is working and what is not, and it addresses energy reduction practices, energy efficiency, and environmental challenges, practical ways to reduce energy consumption, home modifications, and future energy demand. It then outlines why public education is so critical and how it plays an enormous part in the future of the problem.

Climate Management looks at cap-and-trade policies and other forms of *mitigation* that have been proposed by governments worldwide in order to get a handle on this runaway problem. One mitigation technique discussed in detail is the use of *renewable* energy resources, and why solar, wind, geothermal, hydroelectric, and other sources of clean, renewable energy are going to play a critical role in the future.

The volume takes the reader into the exciting world of high-tech computer modeling and the newest advances in analyzing weather and climate via mathematical models to predict future global warming and the impacts of greenhouse gases. It explores the fascinating science of weather prediction, advanced *climate modeling* and computer *simulation* of the atmosphere and ocean, and *remote sensing* and *satellite* data and their roles in long-term monitoring and change detection. It also looks at the latest scientific discoveries, where continued research needs to occur, and technologies waiting on the horizon.

Climate Management expands on government and public involvement, presents readers with numerous activities that everyone can do to combat the problem, and explains how concerned environmentalists can get involved today. Finally, it takes a look into the future and what lies ahead in this warming world.

The Beginning of Global Warming Management

Over the past several decades, technology has progressed to the point where humans can make use of the Earth's natural *resources* to make their lives much more comfortable and enjoy a higher standard of living. However, this has come with a heavy price. As populations have continued to grow, human consumption of natural resources has continued to accelerate, causing severe global environmental problems.

Increasing demands have stressed limited existing resources and contributed to *global warming*. The very health of the *ecosystems* that humans depend on for survival is becoming threatened and endangered. In addition, unsustainable consumption of *fossil fuels* and other resources such as water has not only contributed to global warming, but has also stressed developing countries, affected health, caused economic tension, and threatened national security.

This chapter looks at the beginnings of global warming management—the stage when both the scientific and environmental communities' outcry got to the point that official action was finally taken to combat the Earth's rising temperatures. It discusses the human link to the problem and outlines the role of the United Nations (UN), the Kyoto Protocol, and various political responses. Finally, it reviews the Intergovernmental Panel on Climate Change (IPCC) and its importance.

THE HUMAN LINK

Only 20 percent of the Earth's population lives in industrialized countries, yet they consume nearly 70 percent of the Earth's natural resources. The United States is home to less than 5 percent of the Earth's population, but consumes about 25 percent of its natural resources. The *ecological* "footprint," or signature, that a society leaves on the Earth is significant; not only for the population inhabiting the Earth today, but for all generations to come. Ecological footprints are visible at all scales. Locally, they are visible as populations' daily activities affect the air, water, and land quality of cities and states. It is easy to observe polluting industrial processes, mining processes, and traffic congestion. It is more difficult to see how every person's buying or driving decisions affect the entire world.

Each time someone buys a car that is energy inefficient, chooses not to use sustainable, renewable energy or not to recycle and reduce their waste, does not purchase locally to cut down on transportation costs or not limit the amount of CO_2 *emissions* released to the *atmosphere* because of their activities, they affect the world. Individual choices add up. Each person has the power to make a difference by reducing their personal consumption, buying green products, and using clean, efficient technology. Every individual can be accountable in working toward fixing the problem of global warming.

In order to do this, informed decisions need to be made. The problem involves every person on Earth. Business and government actions are also important. Business can take the lead by purchasing recycled supplies (such as paper), investing in ENERGY STAR equipment (Xerox machines, FAX machines, computers, etc.), remodeling to conserve energy, using fluorescent lighting, shutting off lights in unused rooms, using sinks with automatic water shutoff, monitoring heating and air-conditioning, and offering telecommuting working arrangements (working from home).

An article that appeared in Science*Daily* on February 5, 2007, discussed the IPCC's Fourth Assessment Report (AR4) that attributes human-derived *greenhouse gases* as an overwhelmingly significant contributor to observed global warming. According to Gabriele Hegerl, associate research professor at Duke's Nicholas School of the Environment and Earth Sciences, "We are now seeing, not merely predicting, effects of greenhouse warming on a scale and in ways that were not observable before. When you look at the changes in temperature, circulation, ocean warming, Arctic sea ice reduction, and glacial retreat together, it paints a much clearer picture that external drivers, particularly greenhouse gases, are playing a key role. As a result, we can be much more confident that 20th-century *climate* changes were not just linked to natural variability.

"We've studied improved observations from land, sea, and space, as well as better temperature reconstructions covering the last 1,000 years. By comparing observations against modeled projections, scientists are gaining a better sense of which external climate influences have been important.

"Understanding the observations is what this is really all about. For instance, looking at the patterns of change in 20th-century temperatures, we can now distinguish between changes caused by greenhouse gases, man-made *aerosols,* variability in solar *radiation,* and major volcanic eruptions.

"We now are beginning to understand that changes occur at least partly in response to anthropogenic influences on climate," Hegerl concluded.

According to a *New York Times* article on January 14, 2007, there is already enough overwhelming evidence available today to leave no doubt that human interaction is a large piece of the global warming puzzle. While it is difficult to blame global warming for a specific hurricane or flood or drought or forest fire, it is the collective evidence that points to the distinct anthropogenic influence.

Several trends have been identified, as follows:

- The global average minimum nighttime temperature has risen. Scientists at the National Aeronautics and Space Administration (NASA) believe this is unrelated to the Sun and is linked to the greenhouse gases that hold in heat radiating from the Earth's surface long after the Sun has gone down.
- The stratosphere (a portion of the upper atmosphere) has cooled, which happens when excessive amounts of heat are trapped closer to the Earth's surface. If the change was due to a variation in the Sun's output, both atmospheric layers would change simultaneously.
- There has been a universal warming trend over both the land and the oceans. This removes urbanization as a sole cause.
- Improved *climate models* repeatedly confirm the anthropogenic addition of greenhouse gases to the atmosphere.

THE UNITED NATIONS FRAMEWORK CONVENTION ON CLIMATE CHANGE AND THE KYOTO PROTOCOL

The international policy response to climate change began with the negotiation of the United Nations Framework Convention on Climate Change (UNFCCC), which eventually led the way for the creation and establishment of the Kyoto Protocol—the legal framework for global action to cut greenhouse gas (GHG) emissions.

The UNFCCC

The UNFCCC is an international environmental treaty that was produced at the United Nations Conference on Environment and Development (UNCED). The UNCED is also known as the Earth Summit, the Rio Summit, and Eco '92. It was held in Rio de Janeiro from June 3–14, 1992. The purpose of the treaty was to stabilize greenhouse gas concentrations in the atmosphere in order to prevent global warming. As it was set up, the treaty was nonbinding since it did not set any mandatory limits on greenhouse gas emissions for individual countries.

THE IMPACTS OF WARMING ON THE UNITED STATES AND CANADA

According to the IPCC, the United States and Canada will not escape the effects of global warming. In their report issued on April 6, 2007, they confirm that global warming is already affecting the environment. When the atmospheric temperature rises a little higher—even a few degrees—what may merely be uncomfortable heat now may become dangerous to the point of causing death. This will be felt all the way from Florida and Texas to Alaska and Canada's Northwest Territories.

According to Achim Steiner, executive director of the United Nations Environment Programme (UNEP) "Canada and the United States are, despite being strong economies with the financial power to cope, facing many of the same impacts that are projected for the rest of the world." Chicago and Los Angeles will likely face increasing heat waves. Chicago is expected to see a 25 percent increase in heat waves later this century and dangerously hot days in Los Angeles are projected to increase from a dozen per year to between 44 and 100. North American wood and timber production could suffer huge economic losses of $1 to $2 billion a year during the 21st century if climate change triggers diseases, insect infestations, and wildfires. Groundwater aquifers in Texas, South Dakota, Nebraska, Wyoming, Colorado, Kansas, Oklahoma, and New Mexico could see a lessening of recharge of 20 to 40 percent, causing problems for farmers and population centers.

Winter recreation in eastern North America may disappear by the 2050s, striking a hard blow to the recreation industry. Costs to replenish Florida's beaches with new sand after sea-level rise may cost upward of $9 billion.

The IPCC also cautions that severe storm surges could hit Boston and New York City. Cities that rely on melting snow for water, such as those in the drainage basins of the Rocky Mountains and Sierra Nevada, may experience serious water shortages. In particular, increased tension over water availability will result. As rainfall patterns shift, temperatures rise, and *glaciers* melt around the world, the demand for dwindling supplies of water will likely increase tensions across cultural and political borders.

The IPCC predicts that as temperatures rise summer flows will drastically reduce, leaving huge areas without adequate water. As an example, they report that "A warming of a few degrees by the 2040s is

(continues)

> *(continued)*
> likely to sharply reduce summer flows. As population increases, by then Portland, Oregon, alone will need over 918 million cubic feet (26 million cubic meters) of additional water due to climate change and population growth. The Columbia River's water supply is expected to be much lower, however: about 177 million cubic feet (5 million cubic meters) lower."
>
> The IPCC also warns of storm surges and high tides and predicts that by the 2090s, a one in 500-year flood could be a one in 50-year event in New York City, meaning New York could face serious damage sooner and more frequently.

What the treaty did include was provisions for updates (called protocols) that would set mandatory emission limits. The principal update is the Kyoto Protocol, the international agreement that sets binding targets for 37 industrialized countries and the European community for reducing GHGs.

The UNFCCC was opened for signature on May 9, 1992, and entered into force on March 21, 1994. Its principal objective was "to achieve stabilization of greenhouse gas concentrations in the atmosphere at a low enough level to prevent dangerous anthropogenic interference with the *climate system*."

One of the UNFCCC's first major achievements was that it set into place a "National Greenhouse Gas Inventory," which serves as a tabulation of GHG emissions and removals. All countries that signed the treaty must submit a greenhouse gas record on a regular basis. Nations that signed the treaty are divided into three groups: (1) Annex I countries, which are the industrialized countries; (2) Annex II countries, which are the developed countries that pay for costs of developing countries; and (3) developing countries. The UNFCCC has been ratified by the United States, Canada, France, Germany, the United Kingdom, Aus-

UN Headquarters in New York City *(Nature's Images)*

tralia, Austria, Denmark, Finland, and virtually the entire international community (see Appendix).

The Annex I countries agree to reduce their GHG emissions to levels that are below their 1990 levels. If industries exceed their allotted limits they must buy emission allowances or offset their excesses through a mechanism that is agreed upon by the UNFCCC. The Annex II countries (which are a subgroup of the Annex I countries) also participate as OECD (Organisation for Economic Co-operation and Development) members. The developing countries are not expected to cut back on their *carbon* emissions unless developed countries provide them with the necessary funding and technology to accomplish it. Developing countries can become Annex I countries once they have become developed.

There have been opponents to the treaty who believe that not requiring developing countries to control their emissions is not fair. They feel that all countries should have to reduce emissions equally. Some developing countries have said they cannot afford the costs of compliance. Other countries have countered that, saying that the Stern Review calculates the cost of compliance is actually less than the cost of the consequences of doing nothing.

At the Earth Summit on June 12, 1992, 154 nations signed the UNFCCC, which, when ratified, committed those countries to a voluntary agreement to reduce atmospheric concentrations of greenhouse gases with the goal of "preventing dangerous anthropogenic interference with the Earth's climate system." The actions were targeted mostly at industrialized nations to get them to stabilize their emissions of GHG at 1990 levels by the year 2000. On September 8, 1992, the U.S. Senate Foreign Relations Committee approved the treaty, reporting it through Senate Executive Report 102-55 on October 1, 1992. The Senate then consented to ratification on October 7, 1992. President George Bush signed the instrument of ratification on October 13, 1992, and deposited it with the UN Secretary-General. The treaty became effective on March 21, 1994, once it received the ratification of 50 countries. Since that time, the participating nations meet once a year at the Conference of the Parties in order to assess the progress being made in dealing with climate change. In the mid-1990s, negotiations began on the drafting of the Kyoto Protocol to establish legally binding obligations holding participating, developed countries responsible for reducing their greenhouse gas emissions.

The Kyoto Protocol
The Kyoto Protocol established legally binding commitments for reduction of four principal GHGs: carbon dioxide (CO_2), *methane* (CH_4),

The Beginning of Global Warming Management

The Kyoto Protocol entered into force on February 16, 2005.
(IISD/Earth Negotiations Bulletin)

Nitrous oxide (NO_x), and sulphur hexafluoride (SF_6); and two groups of gases: hydrofluorocarbons (HFCs) and perfluorocarbons (PFCs). These are produced by the Annex I countries (industrialized nations). At the UNFCCC Conference of the Parties held in Kyoto, Japan, in 1997, the Kyoto Protocol was adopted for use after long, intense negotiations. The majority of industrialized nations and some central European countries agreed to the Protocol and it entered into force on February 16, 2005. It is significant that the Protocol was the first legally binding agreement enforcing reductions in greenhouse gas emissions. It called for reductions of 6 to 8 percent below 1990 levels to occur between 2008 and 2012; a time period referred to as the first emissions budget period. At that time, the United States would be required to reduce its total emissions by an average of 7 percent below 1990 levels. Neither Presidents Bill Clinton nor George W. Bush sent the Protocol to

Congress for ratification. The Bush administration completely rejected the Protocol in 2001.

The objective of the Protocol was to stabilize greenhouse gas concentrations so that they remain below a level that causes global warming. There are five principal concepts of the Kyoto Protocol, which are as follows:

- Commitment: The Protocol establishes a legally binding commitment of the Annex I countries to reduce their GHG emissions.
- Implementation: Official policies and measures must be prepared by each participating country concerning how they will meet their objectives. Each country must also implement and use all mechanisms possible to absorb GHGs in order to be awarded credits that would allow for additional emissions.
- The impacts on developing countries will be minimized through the establishment of an *adaptation* fund for climate change. This will facilitate the development and deployment of techniques that can help increase resilience to the impacts of climate change.
- Each country is held responsible for accounting, reporting, and review to ensure they are strictly abiding by the terms of the Protocol. They submit annual emission inventories and national reports at regular intervals.
- Compliance: A compliance committee is established to ensure that individual countries are in strict compliance with their commitments under the Protocol.

One of the provisions of the Protocol is the manner in which it sets up an understanding of responsibility. The UNFCCC agreed to what

(opposite page) This map represents the countries that have signed and ratified the Kyoto Protocol (purple); those that have not expressed a position (beige); and those that have signed, but not ratified (yellow).

The Beginning of Global Warming Management

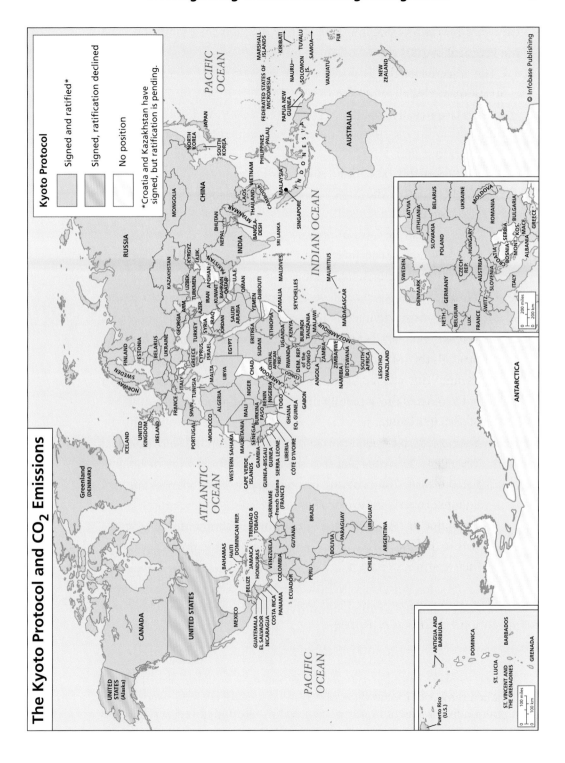

The Kyoto Protocol and CO₂ Emissions

they referred to as "common but differentiated responsibilities." The participating parties agreed on three terms:

1. The developed (industrialized) countries are currently (and have been historically) the largest emitters of GHGs;
2. per capita emissions in developing countries are still relatively low; and
3. the share of global emissions originating in developing countries will grow as their social and developmental needs grow.

What is so critical about this is that China, India, and other developing countries were not included in the original GHG restrictions of the Kyoto Protocol because they were not among the main contributors when the treaty was negotiated. Today, however, both China and India are developed nations. China is developing so rapidly that it is opening an average of one new coal-fired power plant each week, adding enormous amounts of GHGs to the atmosphere, unimpeded by the terms of the Kyoto Protocol. These rapidly developing countries' unaccountability was a principal reason why the Bush administration did not ratify the Kyoto Protocol. Also agreed upon in the original Protocol were financial commitments. It stipulates that it is the responsibility of the developed countries to invest billions of dollars and supply the proper technology to developing countries to finance climate-related studies and projects.

In addition, the Protocol also allows an environmental policy tool called cap and trade (for a more detailed discussion, see chapter 3). What this means is that there are caps (or limits) set on the developed countries (Annex I group) as to how much GHGs they can legally emit. On average, the cap requires countries to reduce their emissions 5.2 percent below their 1990 baseline over the 2008–2012 period. While the caps apply to the country itself, in practicality they are then divided within the country to the various industrial entities—power plants, car and computer manufacturers, and so forth. If a particular industry—a power plant—knows it is going to exceed its allotted quota, it is allowed to purchase credits elsewhere to offset the overage. The purchase of the credits (or excess allowances) are often purchased through a broker or an exchange set up expressly for that—a global carbon market. As a business venture, the Protocol allows groups of Annex I countries to join together to create a market within a market. Several exist today, such as in the European Union (EU), which created the EU Emissions Trading System (EU ETS). The EU ETS uses EAUs (EU allowance units), which are each equivalent to a Kyoto assigned amount unit (AAU). The United Kingdom uses the UK ETS.

The sources of Kyoto credits are what are called the clean development mechanism (CDM) and joint implementation (JI) projects. The CDM allows the creation of new carbon credits by developing emission reduction projects in non–Annex I countries. Under the Protocol, countries' actual emissions have to be monitored and precise records have to be kept of the trades carried out. Registry systems trace and record transactions by countries under the mechanisms. The UN Climate Change Secretariat, based in Bonn, Germany, keeps an international transaction log to verify that transactions are consistent with the rules of the Protocol. The enforcement branch was created and given the responsibility to ensure compliance. If it is determined that an Annex I country is not in compliance with its emissions limitation, then the country is required to make up the difference plus an additional 30 percent. In addition, that country is then suspended from making transfers under an emissions trading program.

Since the Protocol's inception, it has become apparent that in order to meet the original objective of stabilizing GHG emissions to control global warming, even larger emission reductions will need to be achieved than those originally required by Kyoto.

The table on page 14 illustrates the changes in GHG emissions of some prominent countries.

When the United Nations met at their annual climate conference in December 2005 in Montreal, participating nations began negotiations for a second set of targets for the period beginning in 2013 (once the original period ended in 2012). Currently, 2009 is a crucial year in the international arena of finding a workable solution to climate change. In 2007, the parties agreed to create an ambitious and effective international response to climate change to be agreed on at the climate conference in Copenhagen in December 2009.

Greenhouse Gas Emissions of Prominent Countries

COUNTRY	CHANGE IN GHG EMISSIONS (1992–2007)
India	+103%
China	+150%
United States	+20%
Russian Federation	-20%
Japan	+11%
Worldwide Total	+38%

Note: According to estimates from the Netherlands Environmental Assessment Agency (PBL), in the second half of 2008 there was a halving of the annual increase in global CO_2 emissions from fossil fuel use and cement production. Emissions increased by 1.7 percent in 2008 against 3.3 percent in 2007. Since 2002, the overall worldwide annual increase has averaged 4 percent. Besides high oil prices and financial crises, the increased use of renewable energy resources (such as biofuels for highway transportation and wind energy for electricity generation) has caused a noticeable mitigating impact on CO_2 emissions.

CO_2 emissions in the United States fell 3.12 percent in 2008, and, for the first time, were surpassed by those from China. There was a small absolute decline in the European Union as a whole, with declines also reported in Australia and Japan. Emissions in the Eastern European/CIS region increased 1.72 percent in 2008. Emissions from the large developing nations of Brazil, China, and India grew 6.9 percent, 6.6 percent, and 7.2 percent, respectively—together these nations accounted for 27.6 percent of the world total in 2008.

THE U.S. RESPONSE AND INTERNATIONAL REACTIONS

While the bulk of the world's countries agreed to Kyoto, the United States took a different stance, choosing to approach the issue on its own terms.

U.S. Response

The former vice president Al Gore was a main participant in putting the Kyoto Protocol together in 1997. President Bill Clinton signed the agreement on November 12, 1997, but the U.S. Senate refused to ratify it, citing potential damage to the U.S. economy if the nation were forced to comply. The Senate also objected because Kyoto excluded certain developing countries, including China and India, from having to comply with new emission standards.

On March 29, 2001, the Bush administration withdrew the United States from the 1997 Kyoto Protocol on Climate Change. From a statement released by the U.S. Embassy in Vienna, Austria, it said that although the U.S. government was committed to developing an effective way to address the problem of global warming, it believed that the Kyoto Protocol was "fundamentally flawed," and therefore "is not the best approach to achieve a real environmental solution." The administration stated that, "The Kyoto Protocol does not provide the long-term solution the world seeks to the problem of global warming. The goals of the Kyoto Protocol were established not by science, but by political negotiation, and are therefore arbitrary and ineffective in nature. In addition, many countries of the world are completely exempted from the Protocol, such as China and India, who are two of the top five emitters of greenhouse gases in the world. Further, the Protocol could have potentially significant repercussions for the global economy."

President Bush commented on the treaty: "This is a challenge that requires a 100 percent effort; ours, and the rest of the world's. The world's second-largest emitter of greenhouse gases is the People's Republic of China. Yet, China was entirely exempted from the requirements of the Kyoto Protocol. India and Germany are among the top emitters. Yet, India was also exempt from Kyoto . . . America's unwillingness to embrace a flawed treaty should not be read by our friends and allies as any abdication of responsibility. To the contrary, my administration is committed to a leadership role on the issue of climate change . . . Our approach must be consistent with the long-term goal of stabilizing greenhouse gas concentrations in the atmosphere." (Note that as of 2009 China has become the largest GHG emitter in the world.)

Therefore, 10 days after taking office, Bush established a cabinet-level working group to find a more practical method to work with global climate change. The result of the working group was an energy policy that reflected the seriousness of the future of U.S. environmental policy. Bush announced the Clear Skies and Global Climate Change Initiatives in February 2002. The initiatives cover the following goals for managing global climate change:

- By 2018, emissions of the three worst air pollutants will be cut by 70 percent.
- In the next 10 years, the United States will cut greenhouse gas intensity by 18 percent.
- Goals similar to those of the Kyoto Protocol will be achieved, using market-based approaches.

These solutions differ from Kyoto in that they are based on free-market solutions. There are four recommendations:

1. **Ensuring continuing economic growth.** It is in no country's best interest to sacrifice economic growth. With market-based incentive structures to spur innovation, it will be possible to move forward in the field of environmental conservation. Provisions under the Kyoto Protocol would rely on inflexible regulatory structures that would distort investment and waste billions of dollars on pollution permits, accomplishing no real change for the environment.
2. **Finding global solutions.** Addressing this issue must be as comprehensive as possible. All nations including developing countries, must be involved.
3. **Using the most modern technology.** The United States is committed to investing heavily in research and development and encouraging private companies to do the same through market-based incentives. Since 1990, the United States has spent more than all of the countries of the European Union on research in new energy and environmentally friendly technology.
4. **Focusing on bilateral relations to provide assistance.** The United States has already worked with more than 56 countries on their energy and environmental policies.

According to Bush, "The United States fully acknowledges the problem of global warming, and is committed to pursuing a practical and sustainable plan to address this grave situation. The United States hopes to find a workable solution to this serious problem that affects all of us in the global community."

International Reactions

The international reaction to Bush's response to global warming was heated. Although there was faint support from some sectors that the administration finally acknowledged global warming as a problem worthy of attention and committed U.S. involvement toward finding a feasible solution, most reactions were negative. Accusing the administration of trying to create a new ad hoc process—separate from the official framework established by the United Nations—critics stated that the U.S. response would do nothing more than distract from the progress the rest of the world was trying to make toward stabilizing climate change. If anything, they felt it would actually hamper any progress being made to reduce greenhouse gas emissions and slow global warming. Great Britain and Germany especially criticized the United States, stating that all international climate agreements should logically stay within the jurisdiction of the United Nations.

German chancellor Angela Merkel said, "For me, that is nonnegotiable. In a process led by the United Nations, we must create a successor to the Kyoto agreement, which ends in 2012. But it is important that they flow from the United Nations."

Hilary Benn, Britain's international development secretary remarked, "I think it is very important that we stick with the framework we've got. In the end, we have to have one framework for reaching agreement. I think that is very clear."

Leaders from environmental groups also had strong opinions. Philip Clapp, president of the National Environmental Trust, said, "This is a transparent effort to divert attention from the president's refusal to accept any emissions reductions proposals at next week's G8 summit."

David Doniger, the climate policy director for the Natural Resources Defense Council, commented, "There is no more time for longwinded talks about unenforceable long-term goals. We need to get a serious commitment to cut emissions now and in the G8."

The Bush administration offered an alternative environmental plan on June 11, 2001, promising increased environmental research and commitment from the United States. Bush announced that he was "committing the United States of America to work within the United Nations framework and elsewhere to develop an effective science-based response to the issue of global warming."

Bush also stated that, "The rest of the world emits 80 percent of all greenhouse gases, and many of those emissions come from developing countries. The world's second largest emitter of greenhouse gas is China, yet China was entirely exempted from the requirements of the Kyoto Protocol."

Bush committed his administration to fully fund high-priority areas for scientific research into climate change over the next five years and help developing nations to match the U.S. commitment. According to CNN News, former president Clinton signed the Kyoto Protocol but also said he would not send it to the Senate for ratification until several changes were made.

One country that did not seem to be up in arms over the U.S. stand was Australia. The Australian prime minister John Howard supported Bush's plan. According to Howard, "We are a net exporter of energy, and unless you have the developing countries involved we would be hurt. Our position . . . is much closer to that of the United States than the attitude of the European countries. I do think what the president indicates in his speech will lead to an alternative to simply saying "no" to the Kyoto Protocol, and I welcome that."

Pia Ahrenkilde-Hansen, the EU spokeswoman, remarked, "It is positive that the U.S. administration is realizing that there needs to be something done about climate change but we feel that the multilateral approach is the best way to face up to this tremendous challenge."

Many environmental groups opposed Bush's voluntary plan, however, saying that it ultimately would do nothing to curb U.S. emissions. According to a December 4, 2003, *New York Times* report, "The 1997 Protocol had many flaws, but it represented the only international response to the global warming problem thus far devised, and at the very least it provides a plausible framework for collective international action."

The international community was not alone in disagreeing with the Bush administration's stand. Several U.S. cities rose to the occasion and dozens of mayors—representing more than 25 million Americans—pledged that their cities would cut greenhouse gases by 7 percent by 2010.

Greg Nickles, Seattle's mayor who spearheaded the event, says, "This campaign has clearly touched a nerve with the American people. The climate affects Democrats and Republicans alike. Here in Seattle, we rely on the snow for our drinking water and hydroelectricity but it is disappearing."

Nickles also warned that each city had a tough target of cutting emissions by 7 percent, and each mayor would choose a different way to accomplish that goal. He also said, "There are changes we will have to make but there are many opportunities to create employment and make for a better life. In any event, the costs of doing nothing are greater than doing something." Some of the specific proposals for cities include using *hybrid cars,* investing in renewable energy, improving public transportation, planting trees, promoting carpooling, and providing cycling lanes.

The G8

The G8, or Group of Eight, is a forum that was created by France in 1975 for the governments of eight nations of the Northern Hemisphere. The participating members are Canada, France, Germany, Italy, Japan, Russia, the United Kingdom, and the United States. The European Union (EU) is also represented but it cannot host or chair. The table on page 20 lists the current members.

Each year the G8 holds a conference in the country of whoever is currently serving as president. The number of participating countries have evolved over the years since 1975, and just recently it has been proposed that the group be expanded to include five developing countries, referred to as the Outreach Five (O5), which include Brazil, China, India, Mexico, and South Africa. These countries have attended as guests in the past. It has been proposed that the name be changed to the G8+5.

The G8 is an informal forum that began in 1973 after the oil crisis and global recession that followed it. The object of the gathering is to discuss issues of mutual or global concern, such as energy, the environment, terrorism, economics, health, trade, etc. At the Heiligendamm

The G8 Leaders	
COUNTRY	WORLD LEADER
Canada	Prime Minister Stephen Harper
France	President Nicolas Sarkozy
Germany	Chancellor Angela Merkel
Italy	Prime Minister Silvio Berlusconi
Japan	Prime Minister Taro Aso
Russia	President Dimitry Medvedev
United Kingdom	Prime Minister Gordon Brown
United States	President Barack Obama

Summit held in 2007, the G8 addressed the issue of energy efficiency and global warming.

The group agreed, along with the International Energy Agency (IEA), that the best way to promote energy efficiency was on an international basis. As a result, on June 8, 2008, the G8, and China, India, South Korea, and the European Community jointly established the International Partnership for Energy Efficiency Cooperation. The G8 finance ministers agreed to the "G8 Action Plan for Climate Change to Enhance the Engagement of Private and Public Financial Institutions." They also initiated the climate investment funds (CIFs) by the World Bank, which is put into place to help existing efforts until a new framework under the UNFCCC is implemented after 2012, when Kyoto expires.

THE INTERGOVERNMENTAL PANEL ON CLIMATE CHANGE

In order to make meaningful management decisions to minimize the negative impacts of climate change, it is necessary to have an organized body of professionals working together toward the common goal of understanding the science of climate change. This way they

The Beginning of Global Warming Management

can advise political leaders who can then develop regulations that enforce positive human response to that change. The IPCC is a scientific organization established by UNEP and the World Meteorological Organization (WMO) in 1988. The IPCC is comprised of the world's top scientists in all relevant fields who review and analyze scientific studies of climate change and provide authoritative assessments of the state of knowledge regarding global warming. The IPCC was established to provide decision-makers and others interested in climate change with an objective source of information. The IPCC itself does not conduct any research. Its key role is "to assess on a comprehensive, objective, open, and transparent basis the latest scientific, technical, and socio-economic literature produced worldwide relevant to the understanding of the risk of human-induced climate change, its observed and projected impacts and options for adaptation and mitigation." The reports they produce are of a high scientific and technical standard, meant to reflect a range of views and expertise and encompass a wide geographical area.

The IPCC produces reports at regular intervals. To date there have been four major assessments: 1990, 1995, 2001, and 2007. The IPCC is comprised of about 2,500 of the world's top climate scientists and is

Dr. Rajendra Pachauri has been the chair of the IPCC since 2002. He is an environmentalist and also the director general of the Energy and Resources Institute in New Delhi, involved in sustainable development. On December 10, 2007, Dr. Pachauri accepted the Nobel Peace Prize on behalf of the IPCC, along with corecipient Al Gore. *(IISD/Earth Negotiations Bulletin)*

chaired by Dr. Rajendra Pachauri of India. Once the reports are released, they become standard works of reference that are widely used by policymakers, experts, and others. For example, in 1990, the findings of the first First Assessment Report (FAR) played a critical role in establishing the UNFCCC. The Second Assessment Report (SAR), released in 1995, provided key input for the negotiations of the Kyoto Protocol in 1997. The Third Assessment Report (TAR) in 2001 was used in the development of the UNFCCC.

Currently, the IPCC has three working groups and has undertaken the National Greenhouse Gas Inventories Programme (IPCC-NGGIP) in collaboration with the OECD and the IEA. Each working group has its own agenda and is assisted by a technical support unit and the working group or task force bureau. Working Group I (WGI) is titled *The Physical Science Basis*. Working Group II (WGII) is called *Impacts, Adaptation and Vulnerability*. Working Group III (WGIII) is called *Mitigation of Climate Change*.

The main objective of the greenhouse gas inventories programme is to develop and refine a methodology for the calculation and reporting of national GHG emissions and removals. In addition, there is a provision written into the agreement where further task groups and steering groups may be established for a duration of time to consider specific topics or concerns.

Working Group I

WGI assessed the physical scientific aspects of the climate system and climate change. Their latest report, published on February 2, 2007, was released in Paris. This report covers information on changes in greenhouse gases and aerosols in the atmosphere and the role they play in determining the behavior of the climate. The report provides specific details in the changes of air, land, and ocean temperatures, glaciers, rainfall, and ice sheets. It takes into account enormous amounts of satellite-derived data for broad global coverage.

In addition to the current status of the atmosphere, the report also focuses on the past and includes a paleoclimatic review of the Earth's glacial and interglacial periods, the evidence left behind, and how the past can offer clues about the future. This working group also looks at

The IPCC Working Group I speaking about their focus on the Fourth Assessment Report, *The Physical Science Basis,* at the 10th session in Paris, France, on January 29–February 1, 2007. *(IISD/Earth Negotiations Bulletin)*

how climate change interacts and affects geochemistry and the biosphere. Complex climate models are evaluated, and the driving factors—or climate *forcings*—are analyzed so that projections can be made as to what the future climate may be like both globally and locally.

Working Group II

WGII assessed the vulnerability of socioeconomic and natural systems to climate change, the negative and positive consequences of climate change, and options for adapting to climate change. Their most recent report was released on April 6, 2007, in Paris, and was entitled *Impacts, Adaptation and Vulnerability.* It provides a detailed analysis of how global warming is affecting natural and human systems, what its future impacts will be, and to what extent adaptation and mitigation can reduce these impacts. It analyzes how adaptation and mitigation work together and how societies can make the best use of resources they have so that they can maintain a *sustainable development.*

This report looks at specific natural Earth systems, such as ecosystems, water resources, coastal systems, oceans, and forests. It also analyzes human-controlled sectors, such as industry, agriculture, and health. It examines these issues on a geographical basis, breaking the data into subregions such as North America, Latin America, polar regions, Africa, Asia, Australia and New Zealand, Europe, and small islands.

Working Group III

WGIII is responsible for assessing practical options for mitigating climate change through limiting and preventing greenhouse gas emissions. They also focus on identifying methods that remove greenhouse gas emissions from the atmosphere. Their fourth report was released

The IPCC Working Group III focusing on their interest in the Fourth Assessment Report, *Mitigation of Climate Change* at the ninth session in Bangkok, Thailand, on April 30–May 4, 2007. *(IISD/Earth Negotiations Bulletin)*

May 4, 2007, in Bangkok. The report analyzes the world's GHG emission trends and analyzes various mitigation options for the main economic sectors from the present to 2030. It provides an in-depth analysis of the costs and benefits of various mitigation approaches and also looks at short-term strategies and projects how effective they would be in the long term. The report focuses on policy measures and instruments available to governments and industries to mitigate climate change and stresses the strong relationships between mitigation and sustainable development.

The Task Force on National Greenhouse Gas Inventories (TFI) was established by the IPCC to oversee the National Greenhouse Gas Inventories Programme.

IPCC REPORTS

The IPCC's Fourth Assessment Report (AR4), released in 2007, represents the work of more than 1,200 authors and 2,500 scientific expert reviewers from more than 130 countries. The terminology the IPCC uses when they make projections is very specific. When discussing their degree of confidence, the following terminology applies:

Very high confidence	At least a 9 out of 10 chance
High confidence	About an 8 out of 10 chance
Medium confidence	About a 5 out of 10 chance

In terms of likelihood of occurrence:

Extremely likely	> 95 percent
Very likely	> 90 percent
Likely	> 66 percent
More likely than not	> 50 percent
Less likely than not	< 50 percent
Unlikely	> 33 percent
Very unlikely	> 10 percent
Extremely unlikely	> 5 percent

Working Group I Report—*The Physical Science Basis*

This report contains the strongest language yet of any of the IPCC's reports, and it found that it is very likely (> 90 percent probability) that emissions of heat-trapping gases from human activities have caused "most of the observed increase in globally averaged temperatures since the mid-20th century." The report concludes that it is "unequivocal" that Earth's climate is warming, "as is now evident from observations of increases in global average air and ocean temperatures, widespread melting of snow and ice, and rising global mean sea level."

The report also verifies that the current atmospheric concentration of CO_2 and methane "exceeds by far the natural range over the last 650,000 years." Since the beginning of the industrial revolution, concentrations of both gases have increased at a rate that is "very likely to have been unprecedented in more than 10,000 years."

The report also identified the following findings:

- Eleven of the last 12 years were among the 12 hottest years on record.
- Over the past 50 years, cold days, cold nights, and frost have become less frequent, while hot days, hot nights, and heat waves have become more frequent.
- The intensity of hurricanes in the North Atlantic has increased over the past 30 years, which correlates with increases in *tropical* sea surface temperatures. They are likely to become more intense.
- Between 1900 and 2005, the Sahel, the Mediterranean, southern Africa, and parts of southern Asia have become drier, adding stress to water resources in these regions.
- Droughts have become longer and more intense and have affected larger areas since the 1970s, especially in the Tropics and subtropics.
- Since 1990, the Northern Hemisphere has lost 7 percent of the maximum area covered by seasonally frozen ground.
- Mountain glaciers and snow cover have declined worldwide.
- Satellite data since 1978 show that the extent of Arctic sea ice during the summer has shrunk by more than 20 percent.
- Since 1961, the world's oceans have been absorbing more than 80 percent of the heat added to the climate, caus-

ing ocean water to expand and contributing to rising sea levels.
- If no action is taken to reduce emissions, the IPCC concludes that there will be twice as much warming over the next two decades than if the GHGs had been stabilized at their 2000 levels.
- The full range of projected temperature increase has now been revised to 2–11.5°F (1.1–6.4°C) by the end of the century because higher temperatures reduce the amount of CO_2 that the land and ocean can hold, keeping more stored in the atmosphere.
- Warming is expected to be greatest over land and at most high northern latitudes and least over the Southern Ocean and parts of the North Atlantic Ocean.
- High latitude precipitation will increase, and subtropical lands (e.g., Egypt) will face drought.
- Extreme heat, heat waves, and heavy precipitation will become more frequent.
- Sea ice is projected to shrink in both the Arctic and Antarctic under all model simulations. Some projections show that by the latter part of the century, late-summer Arctic sea ice will disappear almost entirely.
- Increasing atmospheric CO_2 concentrations will lead to increasing acidification of the oceans, destroying coral and other fragile marine ecosystems.

The IPCC also states that it is very likely that the Atlantic Ocean conveyor belt will be 25 percent slower on average by 2100 (with a range from 0 to 50 percent). Nevertheless, Atlantic regional temperatures are projected to rise overall due to more significant warming from increases in heat-trapping emissions. The models used by the IPCC project that by the end of this century, the global average sea level will rise between 7–23 inches (17–58 cm) above the 1980–1999 average. In addition, recent observations show that meltwater can run down cracks in the ice and lubricate the bottom of ice sheets, resulting in faster ice flow and increased movement of large ice chunks into the ocean, contributing to sea-level rise.

Working Group II Report—*Impacts, Adaptation, and Vulnerability*

WGII describes global warming's effects on society and the natural environment and some of the options available for adapting to these effects. The IPCC has determined that anthropogenic warming over recent decades is already affecting many physical and biological processes on every continent. Of the 29,000 observational pieces of data reviewed, almost 90 percent showed changes that were consistent with the response expected of global warming. In addition, the observed physical and biological responses have been the greatest in the regions that warmed the most.

The major conclusions stated in this report include the following:

- Hundreds of millions of people face water shortages that will worsen as temperatures rise. The most at risk are regions currently affected by drought, areas with heavily used water resources, and areas that get their water from glaciers and snowpack such as the western United States.
- The land area affected by drought is expected to increase, and water resources in affected areas could decline as much as 30 percent by midcentury. U.S. crops that are already near the upper end of their temperature tolerance range or depend on strained water resources could suffer with further warming.
- More than one-sixth of the world's population currently lives near rivers that derive their water from glaciers and snow cover; these communities can expect to see their water resources decline over this century.
- Melting glaciers in areas like the Himalayas will increase flooding and rockslides, while flash floods could increase in northern, central, and eastern Europe.
- The IPCC expects food production to decline in low-latitude regions (near the equator), particularly in the seasonally dry Tropics, as even small temperature increases decrease crop yields in these areas.
- The IPCC projections show drought-prone areas of Africa to be particularly vulnerable to food shortages due to a reduc-

tion in the land area suitable for agriculture; some rain-fed crop yields could decline as much as 50 percent by 2020.
- Under local average temperature increases, regions such as northern Europe, North America, New Zealand, and parts of Latin America could benefit from increased growing season length, more precipitation, and/or less frost, depending on the crop. However, these regions may also expect more flooding. In addition, depending on existing soil types, agriculture may or may not even be feasible.
- Up to 30 percent of plant and animal species could face extinction if the global average temperature rises more than 3–5°F (1.5–2.5°C) relative to the 1980–1999 period. Many say the low range could be reached by midcentury.
- Spring has been arriving earlier during this time, influencing the timing of bird and fish migration, egg laying, leaf unfolding, and spring planting for agriculture and forestry. It can threaten and endanger species by altering the timing of migration, nesting, and food availability, causing them to be out of sync.
- Many species and ecosystems may not be able to adapt to the effects of global warming and its associated disturbances (including floods, drought, wildfire, and insects), causing mass extinctions.
- Experts expect coral reefs and mangroves in Africa to be degraded to the point that fisheries and tourism suffer.
- Some areas, such as the national parks of Australia and New Zealand and many parts of tropical Latin America, are likely to experience a significant loss of *biodiversity*.
- Flooding caused by sea-level rise is expected to affect millions of additional people every year by the end of this century, with small islands and the crowded delta regions around large Asian rivers (such as the Ganges-Brahmaputra) facing the highest risk.
- Regions especially at risk are low-lying areas of North America, Latin America, Africa, the popular coastal cities of Europe, crowded delta regions of Asia that face flood risks

from both large rivers and ocean storms, and many small islands (such as those in the Caribbean and South Pacific) whose very existence is threatened by rising seas.
- Scientists expect heat waves, droughts, wildfires, floods, severe storms, and dust transported between continents to cause locally severe economic damage and substantial social and cultural disruption. The IPCC projects an extended fire season for North America as well as increased threats from pests and disease.
- In cities that experience severe heat waves, scientists project an increase in the incidence of cardiorespiratory diseases caused by the higher concentrations of ground-level *ozone* (smog) that may accompany higher air temperatures. Some infectious diseases, such as those carried by insects and rodents, may also become more common in regions where those diseases are not currently prevalent (such as dengue fever, malaria, yellow fever, encephalitis, lyme disease, and visceral leishmaniasis).
- Many of the unavoidable near-term consequences of global warming can be addressed through adaptation strategies such as building levees and restoring wetlands to protect coasts, altering farm practices to grow crops that can survive higher temperatures, building infrastructure that can withstand extreme *weather,* and implementing public health programs to help people in cities survive brutal heat waves. This is a more serious problem, however, for developing countries that lack the economic wherewithal to build appropriate infrastructure.

Working Group III—*Mitigation of Climate Change*

There are several strategies available today that the IPCC believes could slow global warming and prevent the worst environmental consequences if they were implemented immediately. While there has been some criticism that implementing proper measures to halt global warming would be too expensive, the IPCC has determined that the economic impact on the world economy would only be a fraction of a percent reduction in the annual average growth rate of global *gross domestic product* (*GDP*).

The IPCC also warns that the policies that have been put into place so far have not been robust enough to stop the growth of global emissions caused by the increased use of fossil fuels, *deforestation,* overpopulation, and wildfires. It is critical that clean technologies are developed in order to reduce emissions and stop global warming. Although there has been much talk about reducing emissions, there has been an increase in heat-trapping gases of 70 percent from 1970 to 2004. Of these, CO_2 emissions account for 75 percent of the total *anthropogenic emissions*. The emission growth rate is expected to continue if serious changes are not made immediately.

In 2004, developed countries (such as the United States) had 20 percent of the world population and contributed nearly three-quarters of the global emissions. Developing countries generated only one-quarter of the emissions. The IPCC has projected that CO_2 emissions from energy use are projected to increase 45 to 110 percent if fossil fuels continue to dominate energy production through 2030, with up to three-fourths of future emission increases coming from developing countries (such as China and India).

The IPCC analyzed several mitigation options—some of them efficient enough to bring about a 50–85 percent reduction in emissions of greenhouse gases by 2050 (compared with 2000 levels). Predictions with these models put GHG concentrations at the end of the century at 445–490 ppm. As a comparison, the IPCC says if mitigation of this nature does not take place and GHG levels continue to increase, concentration levels could reach 855–1,130 ppm. The IPCC believes there will be more mitigation technologies available before 2030 that could lead to even greater emissions reductions. They believe that the search for energy efficiency will play a key role in the future and support larger investments in research and development to stimulate deployment of new technological advances. They also stress the importance of increasing government funding for research, development, and demonstration of carbon-free energy sources.

The U.S. Political Arena

In order to get global warming effectively under control, it will take the efforts of every country worldwide. Because of the immensity of the issue, the backing of national governments is critical—legislatively and economically. This chapter discusses the current political climate in the United States and Washington's stand on the global warming issue, including a personal look at President Obama's view on global warming. Next, it examines the connection global warming has with national security and terrorism and what the nation could expect if the problem is not brought under control. Finally, it presents the current legislation being considered in the United States.

THE CURRENT POLITICAL CLIMATE

Historically, the United States has not been a leader in stressing the importance of the global warming issue. According to "The One Environmental Issue," a January 1, 2008, *New York Times* editorial, when Al

Gore ran for president in 2000 he could have made the global warming issue a key point in his campaign, but his advisers persuaded him that it was too complicated and forbidding an issue to sell to ordinary voters. John Kerry's ideas for addressing climate change and broaching the idea of lessening the nation's dependence on foreign sources of oil made no headway either.

Although some politicians have tried to get involved in environmental issues, the overall trend has been one of inaction. However, times seem to be changing. Severe weather events are occurring, species are becoming endangered, glaciers are melting, and areas are suffering from drought. The media has finally taken on the role of making the public aware of the effects of a warming world. The big question still remains to be answered, however: To what extent are Americans willing to accept responsibility for the threat, take action, and make the personal sacrifices necessary to control the problem? To be specific—are Americans finally willing to pay slightly more for alternate, renewable energy and significantly change their lifestyles in order to reduce the use of fossil fuels?

Even though Al Gore did not focus on global warming during his campaign, he has had phenomenal influence since and played a critical role in educating the public about the issue and why it has to be dealt with now. His film and book, *An Inconvenient Truth,* have made the public well aware of the issue. So much so, in fact, that survey polls show that the American population is becoming increasingly alarmed. In 2007, the Intergovernmental Panel on Climate Change (IPCC) and Al Gore shared the Nobel Peace Prize for their efforts to bring the issue to the world's attention.

One thing that has frustrated many Americans is that the U.S. government—typically a leader in global issues—has seemed to move so slowly to take action to halt the emissions of greenhouse gases (GHG). State governments are not holding back and waiting any longer. Governors from half of the states have put into effect agreements to lower GHG. Even federal courts have ordered the executive branch to start regulating GHGs. Currently, the Senate is working on a bipartisan bill that would reduce emissions by almost 65 percent by 2050.

Many opponents of the U.S. stance on global warming over the years have openly criticized the lack of federal coordination and action. As more people become aware of the issues involved in global warming, more pressure is being applied in the political arena to take action to slow the process before irreparable damage is done. *(Nature's Images)*

During the 2008 presidential campaign and election, environmental issues did become important talking points. John McCain—who had encouraged taking positive action to fight global warming all along—was serious about dealing with the issue. In 2003, with Joseph Lieberman, Senator McCain introduced the first Senate bill aimed at mandatory reductions in emissions of 65 percent by midcentury. In the Democratic race, all of the original candidates promised that major investments would be made in cleaner fuels and delivery systems, including underground carbon storage for coal-fired plants. They also promised efforts to work toward a new international agreement to replace the Kyoto Protocol when it expired in 2012.

In a *New York Times* article on April 1, 2009, entitled "Democrats Unveil Climate Bill," a new bill to stop heat-trapping gases and wean the United States off foreign sources of oil was announced. The bill has not gained Republican support yet, meaning it will take longer to work its way through Congress. The bill, written by Representatives Henry A. Waxman (D-CA) and Edward J. Markey (D-MA), sets an ambitious goal for capping heat-trapping gases—even higher than President Obama's initial plan. The bill requires that emissions be reduced 20 percent from 2005 levels by 2020 (Obama's called for a 14-percent reduction over the same time period). Both proposals would reduce GHGs by about 80 percent by 2050.

The Waxman-Markey bill, H.R. 2454: American Clean Energy and Security Act of 2009, would require the nation to produce one-fourth of its electricity from renewable energy sources, such as solar, wind, or geothermal by 2025. It also calls for a modernization of the nation's electric grid, production of more electric vehicles, and major increases in energy efficiency in buildings, appliances, and the generation of electricity.

What the proposal does not address, is how pollution allowances would be distributed or what percentage would be auctioned off or given for free. It also does not address how the majority of the billions of dollars raised from pollution permits would be spent or whether the revenue would be returned to consumers to compensate for higher energy bills. These are some of the issues Congress will need to address.

Under Obama's plan, about 65 percent of the revenue from pollution permit actions would be returned to the public in tax breaks. Several members of Congress would like to see all the revenue from any carbon reduction plan returned to the public.

Mr. Waxman, who serves as the chairman of the Energy and Commerce Committee, said that his measure would create jobs and provide a gradual transition to a more efficient economy. "Our goal is to strengthen our economy by making America the world leader in new clean-energy and energy-efficiency technologies."

For coal-producing states, the bill offers $10 billion in new financing for the development of technology to capture and store emissions of CO_2 from the burning of coal. A coalition of business and environmental groups, United States Climate Action Partnership, said the measure is a "strong starting point" for addressing emissions of heat-trapping gases and that it had incorporated many of the partnership's recommendations.

PRESIDENT OBAMA AND HIS OUTLOOK ON GLOBAL WARMING

On January 20, 2009, when Barack Obama was sworn in as the 44th president of the United States, he delivered a speech after taking the oath of office. In it, he stressed that "Each day brings further evidence that the ways we use energy strengthen our adversaries and threaten

President Barack Obama has taken a stand to address global warming. Current plans include controlling GHG emissions, helping American automakers produce more environmentally friendly cars and reducing the country's dependence on foreign oil by turning instead to renewable energy sources. *(U.S. Embassy)*

our planet." He also affirmed that the energy challenges the nation faces today are a very real crisis that must be dealt with, and he promised a waiting nation that "we will harness the Sun and the winds and the soil to fuel our cars and run our factories . . . in an effort to roll back the specter of a warming planet." He also promised that the nation would no longer "consume the world's resources without regard to effect."

Prior to his inauguration address, Obama had sent a video message to an international summit meeting on global warming organized by Governor Arnold Schwarzenegger of California, held in Beverly Hills, California, on November 18–19, 2008. Obama stressed that despite the continuing economic turmoil, reductions in GHG emissions would remain a central component of his energy, environmental, and economic policies. The message he sent was clear. The need to curb heat-trapping gases will be a priority for his administration. He also stressed that the energy revolution the nation could expect from his administration would overcome what he called America's "shock and trance" cycle as oil prices spike and collapse. The following is his explanation of the shock-and-trance cycle (taken from the CBS transcript of *60 Minutes* on November 16, 2008):

> *Steve Kroft*: When the price of oil was at $147 a barrel, there were a lot of spirited and profitable discussions that were held on energy independence. Now you've got the price of oil under $60.
>
> *Mr. Obama*: Right.
>
> *Mr. Kroft*: Does doing something about energy, is it less important now than . . . ?
>
> *Mr. Obama*: It's more important. It may be a little harder politically, but it's more important.
>
> *Mr. Kroft*: Why?
>
> *Mr. Obama*: Well, because this has been our pattern. We go from shock to trance. You know, oil prices go up, gas prices at the pump go up, and everybody goes into a flurry of activity. And then the prices go back down and suddenly we act like it's not important, and we start, you know, filling up our SUVs again. And, as a consequence, we

never make any progress. It's part of the addiction, all right. That has to be broken. Now is the time to break it.

The following is a transcript of the video message President Obama sent to Schwarzenegger at the summit meeting on global warming (taken from Revkin, *New York Times,* November 18, 2008):

Few challenges facing America—and the world—are more urgent than combating climate change. The science is beyond dispute and the facts are clear. Sea levels are rising. Coastlines are shrinking. We've seen record drought, spreading famine, and storms that are growing stronger with each passing hurricane season. Climate change and our dependence on foreign oil, if left unaddressed, will continue to weaken our economy and threaten our national security. I know many of you are working to confront this challenge. We've also seen a number of businesses doing their part by investing in clean energy technologies. Too often, Washington has failed to show the same kind of leadership. My presidency will mark a new chapter in America's leadership on climate change that will strengthen our security and create millions of new jobs in the process. That will start with a federal cap and trade system. We will establish strong annual targets that set us on a course to reduce emissions to their 1990 levels by 2020 and reduce them an additional 80 percent by 2050. We will invest in solar power, wind power, and next-generation *biofuels.* The United States cannot meet this challenge alone. Solving this problem will require all of us working together. I look forward to working with all nations to meet this challenge in the coming years. Now is the time to confront this challenge once and for all. Delay is no longer an option. Denial is no longer an acceptable response. The stakes are too high. The consequences, too serious. Stopping climate change won't be easy. It won't happen overnight. But I promise you this: When I am president, any governor who's willing to promote clean energy will have a partner in the White House. Any company that's willing to invest in clean energy will have an ally in Washington. And any nation that's willing to join the cause of combating climate change will have an ally in the United States of America.

Then, in a political presentation given on January 26, 2009, President Obama delivered a speech concerning jobs, energy, and climate change, during which he made the following points about his policy on global warming:

- Year after year, decade after decade, we've chosen delay over decisive action. Rigid ideology has overruled sound science. Special interests have overshadowed common sense. Rhetoric has not led to the hard work needed to achieve results and our leaders raise their voices each time there's a spike on gas prices, only to grow quiet when the price falls at the pump.
- Now America has arrived at a crossroads. Embedded in American soil, in the wind and the Sun, we have the resources to change. Our scientists, businesses, and workers have the capacity to move us forward.
- It falls on us to choose whether to risk the peril that comes with our current course or to seize the promise of energy independence. And for the sake of our security, our economy and our planet, we must have the courage and commitment to change.
- It will be the policy of my administration to reverse our dependence on foreign oil while building a new energy economy that will create millions of jobs.
- Today I'm announcing the first steps on our journey toward energy independence, as we develop new energy, set new fuel efficiency standards and address greenhouse gas emissions.
- We will make it clear to the world that America is ready to lead. To protect our climate and our collective security, we must call together a truly global coalition. I've made it clear that we will act, but so too must the world. That's how we will deny leverage to dictators and dollars to terrorists, and that's how we will ensure that nations like China and India are doing their part, just as we are now willing to do ours.
- We have made our choice: America will not be held hostage to dwindling resources, hostile regimes, and a warming planet. We will not be put off from action because action is hard. Now is the time to make the tough choices. Now is the time to meet the challenge at this crossroad of history by choosing a future that is safer for our country, prosperous for our planet, and sustainable.

Obama stressed that the federal government must work with, not against, the individual states to control global warming. His plan also outlined the goal of requiring cars to meet a 35 MPG fuel efficiency standard by 2020 and vowed to "help the American automakers prepare for the future, build the cars of tomorrow, and no longer ignore facts or science." Global warming is real, and his energy policy will be dictated to deal with global warming and will free U.S. dependence on foreign oil for security purposes.

NATIONAL SECURITY AND TERRORISM

In October 2003, Andrew Marshall, a highly respected U.S. Department of Defense (DoD) planner, commissioned a Pentagon study on climate change and U.S. security. The study's principal authors were Doug Randall of the Global Business Network (a California think tank) and Peter Schwartz, former head of planning for Shell Oil. Their conclusion was that global warming could ultimately prove to be a greater risk to the nation than terrorism.

Randall and Schwartz, who interviewed leading climate change scientists, conducted additional research, and reviewed numerous climate models with experts in climatology, concluded that global warming could lead to a slowing of ocean currents. Major currents in the ocean carry huge amounts of heat from the equator to the poles, circulating heat energy on the surface and at great depths. One extremely important current moves in a winding, endless loop; scientists refer to its conveyor belt–like properties as the thermohaline circulation (THC). This global current is significant to major parts of the world—it moves the warm salty Atlantic water that originates near the equator northward toward Greenland and Labrador, where it then cools and sinks. The current sinks more than one mile (1.6 km) in specific locations, where it then turns over and heads south making its way back through the Atlantic toward the equator again. From there, the water continues to move south, travels around the southern tip of Africa, and rises to the surface in the Indian and Pacific Oceans, as well as areas near Antarctica. It then heads north toward the equator again, where it picks up heat, and repeats the cycle.

The problem with adding large amounts of freshwater to the ocean through the melting of ice caps and glaciers is that it decreases the salinity of the ocean water and slows the overturning process at the high

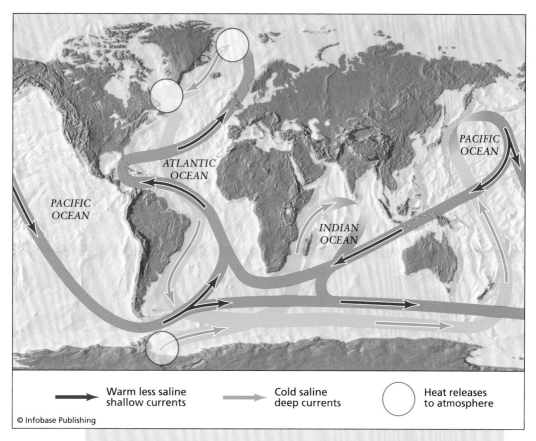

Working as a massive conveyor belt of heat, the ocean thermohaline circulation has a significant effect on weather worldwide. As global warming continues to heat up the planet, many scientists are worried that the addition of freshwater to the ocean from the melting of the Greenland ice sheet could stop the North Atlantic conveyor. If it did shut down, or even slow down, it would send colder temperatures to Europe and cause other sudden climate changes around the world.

latitudes. By slowing the process, it slows down the entire conveyor belt, which means that warmth from the equator will not be brought up into the Northern Hemisphere.

The Gulf Stream, which is the current that transports a significant amount of heat northward from the Earth's equatorial region toward western Europe helping to warm its climate, is part of that circulation system. In fact, if it were not for the Gulf Stream, the North Atlantic and Europe would be on average 9°F (5°C) cooler. If this extensive current were to shut down, it would have a negative impact on the entire ocean/atmospheric system and cause adverse effects worldwide not only in ocean circulation, but also in the jet stream in the atmosphere that drives storm systems. Based on evidence retrieved from ice cores in Greenland, scientists have determined that the THC has been shut down in the past and that every time it has been shut down, an abrupt climate change has occurred. The chief mechanism for shutting down the THC is the addition of freshwater.

The report goes on to analyze how an abrupt climate change scenario could "potentially de-stabilize the geopolitical environment, leading to skirmishes, battles, and even war due to resource constraints such as:

1. Food shortages due to decreases in net global agricultural production;
2. Decreased availability and quality of freshwater in key regions due to shifted precipitation patterns, causing more frequent floods and drought;
3. Disrupted access to energy supplies due to extensive sea ice and storminess."

As these conditions persist and global and local carrying capacities are reduced, tensions could mount around the world, leading to two principal strategies: defensive and offensive. Nations that have the resources and are in a position to do so may build fortresses around their countries, protecting and keeping the resources for themselves. Less fortunate nations—especially those who share borders with warring nations—may engage in battle for access to food, clean water, or energy. Unlikely alliances could be formed as defense priorities shift, and the goal becomes resources for survival instead of religion, ideology, or national honor.

If these chains of events were to occur, it would pose new challenges for the United States. Randall and Schwartz suggest that in order to be prepared to deal with such changes, it is important that the United States:

- improve predictive climate models to allow investigation of a wider range of possible scenarios in order to be able to anticipate how and where changes could happen;
- determine potential impacts of abrupt climate change, through modeling, and how it could influence food, water, and energy;
- determine which countries are most vulnerable to climate change and could contribute materially to an increasingly disorderly and potentially violent world;
- Identify "no-regrets" strategies such as enhancing capabilities for water management;
- Rehearse adaptive responses;
- Explore local implications;
- Explore geoengineering options that control the climate.

The authors advised the DoD to look at potential responses now because there is already evidence in place that global warming has reached a threshold where the THC could start to be significantly affected, such as documented measurements of the North Atlantic being freshened by melting glaciers, increased precipitation, and increased freshwater runoff making it substantially less salty over the past 40 years. Because of this, Randall and Schwartz recommend the report be elevated from a scientific debate to a U.S. national security concern. In their research, they concluded that weather-related events can have an enormous impact on society. They influence food supply, conditions in cities, availability and access of clean water, and the availability of energy.

According to the Climate Action Network of Australia, climate change will probably reduce rainfall in rangeland areas, which would cause a 15-percent drop in grass productivity. This could cause a reduction of the average weight of cattle by about 12 percent, which would significantly reduce the world beef supply. In addition, dairy cows would probably produce 30 percent less milk and insects may invade new fruit-growing areas. Drinking-water supplies would also be affected, possibly causing a 10-percent reduction in water supply. With this given scenario, several major food-producing regions around the world over the next 15 to 30 years may not be able to meet demand.

When population numbers are added to the equation, the situation becomes dire. Currently, more than 400 million people live in the dry, subtropical, overpopulated, and economically poor regions where the negative effect of global warming poses a severe risk to their political, economic, and social stability. In other countries that completely lack resources, the situation will be even worse. In these countries, it is expected that there will be mass emigration as desperate people seek better lives in regions such as the United States that have the resources available to allow them to adapt. This scenario has immediate implications for issues concerning food supply, health and disease, commerce and trade, and their consequences for national security. What the study concluded was that large population movements are inevitable. Learning how to manage populations and border tensions will be critical, and new forms of security agreements dealing specifically with energy, food, and water will be needed. Disruption and conflict will become an everyday way of life.

CURRENT LEGISLATION

The ultimate goal of political action on climate change is to limit and/or reduce the concentration of GHGs in the atmosphere. Political action is a critical component necessary to make any significant global change because without the implementation of the necessary laws and regulations—such as GHG emissions limits, regulatory frameworks within which carbon trading markets can operate, reportable and trackable systems of accountability, and tax incentives or funding assistance—productive and long-term change is not feasible.

Although the United States had a slow start toward addressing the global warming issue, current legislation is now percolating, and progress is slowly being made. The global warming issue has also made it to the Supreme Court. On April 2, 2007, in one of its most important environmental decisions in years, the U.S. Supreme Court ruled that the EPA now has the authority to regulate heat-trapping gases in automobile emissions. The Court further stipulated that the EPA could in no manner "sidestep its authority to regulate the greenhouse gases that contribute to global climate change unless it could provide a scientific basis for its refusal." This gives the EPA the right to regulate carbon dioxide (CO_2) and other heat-trapping gases under the Clean Air Act.

According to Justice John Paul Stevens, "The only way the agency could avoid taking further action now was if it determined that greenhouse gases do not contribute to climate change or provides a good explanation why it cannot or will not find out whether they do."

The Supreme Court also heard another case concerning the Clean Air Act, giving the EPA a broader authority over factories and power plants that want to expand or increase their emissions of air pollutants. Under this broader reading, they made a ruling of 9 to 0 against the Duke Energy Corporation of North Carolina in favor of the EPA, which made environmentalists ecstatic, marking a historic occurrence in the U.S. Supreme Court as a positive step toward the mitigation of global warming. Interestingly, since the ruling on the first case, there has been a growing interest among various industrial groups in working with environmental organizations on proposals for emissions limits.

According to a *New York Times* article on April 3, 2007, Dave McCurdy, president of the Alliance of Automobile Manufacturers, said in response to the decision that, "The Alliance looks forward to working constructively with both Congress and the administration in addressing this issue. This decision says that the EPA will be part of this process."

Although many claimed victory with the Supreme Court's decision, not everyone was satisfied. Chief Justice John G. Roberts, Jr., believed the court should never have addressed the question of the agency's legal obligations in the first place.

On April 17, 2009, the EPA formally declared CO_2 and five other GHGs to be pollutants that endanger public health and welfare. This landmark decision will now put in motion a process that will lead to the regulation of GHGs for the first time in U.S. history. According to the EPA, "The science supporting the proposed endangerment finding was compelling and overwhelming." The decision received diverse reactions. Many Republicans in Congress and industry spokesmen warned that regulation of CO_2 emissions would raise energy costs and kill jobs. Democrats and environmental advocates, however, said the decision was long overdue and would bring long-term social and economic benefits.

Lisa P. Jackson, the EPA administrator, said, "This finding confirms that greenhouse gas pollution is a serious problem now and for future generations. Fortunately, it follows President Obama's call for a low-carbon economy and strong leadership in Congress on clean energy and climate legislation."

The ruling will be followed by a grace period for comments to be made and legislation to emerge from Congress. Once this has occurred, the EPA will determine specific targets for reductions of heat-trapping gases and new requirements for energy efficiency in vehicles, power plants, and industry. At that point, the EPA will begin the process of regulating the climate-altering substances under the Clean Air Act.

A *New York Times* article of December 18, 2007, stated that the Congress plans to create a huge new industry with the purpose of converting agricultural wastes and other plant material into fuel, citing as its primary motive the reduction of the nation's dependence on foreign sources of oil and the cutting back of greenhouse gas generation. What Congress is proposing has far-reaching objectives—the fuel types proposed have not been produced commercially in the United States before and not everyone backs the idea. Some critics claim the technology is immature, the economics are uncertain, hundreds of new factories will be required, and a huge capital investment will be necessary.

According to Mark Flannery, head of energy equity research at Credit Suisse, when asked about the plan's feasibility: "It's not clear that it is doable, but it wasn't clear you could send a man to the moon, either. You don't know until you try."

Historically, Washington's efforts in finding new solutions to energy demand and efficiency were to develop more fuel-efficient cars, not alternative-fuel cars, making this new approach by Congress significant. Other portions of the bill are equally groundbreaking. The bill calls for a significant increase in the amount of ethanol used in the nation's fuel supply. Congress is proposing to double the nation's current level of production to 15 billion gallons (57 billion l). It also foresees that by 2022, an additional 21 billion gallons (79 billion l) a year of ethanol or other biofuels will be produced by developing technology that can obtain useful energy from *biomass* such as straw, tree trimmings, corn stubble, and even common garbage.

Another reason why political involvement is crucial is that in order to accomplish these goals, the nation's key scientists and business leaders will need political and financial support to successfully deal with the technical, environmental, and logistical obstacles they will encounter.

Martin Keller, the director of the Department of Energy (DOE) BioEnergy Science Center at the Oak Ridge National Laboratory in Tennessee, said, "We have the opportunity to revolutionize the way we create fuel for transportation. If we focus on this, we can replace between 30–50 percent of our gasoline consumption with new biofuels."

Christopher G. Standlee, executive vice president of Abengoa Bioenergy remarked, "It certainly is a challenge, but an achieveable challenge."

Under the new legislation, corn ethanol use would reach 15 billion gallons (57 billion l) by 2015. Mandates for next-generation biofuel use would reach 9 billion gallons (34 billion l) in 2017 and 21 billion gallons (79 billion l) by 2022. The bill does contain an escape clause, allowing the government to modify the mandates if they do not prove feasible.

The measure is not without uncertainty or critics. Some have expressed concern at the short time line of only five to 15 years. According to Aaron Brady, an ethanol expert at Cambridge Energy Research Associates, "Congress is making the assumption that the technology will appear. To make billions of gallons of next-generation biofuels, a lot of things have to go right within the space of only a few years."

Brady estimates that more than 100 additional corn ethanol plants will be required, along with at least 200 other biomass fuel plants, a number that could rise depending on how technology develops. He also figures that 700,000 tons (635,000 metric tons) of biomass would be needed each year for a distillery to produce 50 million gallons (189 million l) of ethanol, which adds up in energy costs to transport it.

Some environmentalists remain uneasy because ethanol produced from corn still requires energy and fertilizer involving the use of natural gas, oil, and coal. Some food producers argue that the plan would require growing 20 million more acres (8 million ha) of corn—leaving fewer farming acres for fruits, vegetables, soybeans, alfalfa, and other crops and leading to higher food prices.

As with all important issues, there are always pros and cons that must be taken into account when making decisions. To date, there are a number of congressional acts, bills, and legislative proposals concerning the global warming issue. Some of them are summarized below.

Global Warming Pollution Reduction Act of 2007

The Global Warming Pollution Reduction Act of 2007 (S.309), also known as the Sanders-Boxer bill, was proposed as a bill to amend the Clean Air Act to reduce emissions of CO_2. Introduced in the 110th Congress by Senators Bernie Sanders (I-VT) and Barbara Boxer (D-CA) on January 15, 2007, it was based on the increasing scientific evidence that "global warming is a serious threat to both the national security and economy of the United States, to public health and welfare, and to the global environment; and that action can and must be taken soon to begin the process of reducing emissions substantially over the next 50 years." The bill is considered the most aggressive bill on global warming and is backed by former vice president Al Gore.

The bill listed several targets, incentives, and requirements that the EPA would employ to reduce emissions and help stabilize global concentrations of GHGs. The bill set a goal of reducing U.S. greenhouse gas emissions to a stable global concentration below 450 ppm—a level advised by leading global warming scientists. It required the United States to reduce its emissions to 1990 levels by 2020 and make additional reductions between 2020 and 2050. Specifically, by 2030, the United States would have to reduce its emissions by one-third of 80 percent below 1990 levels; by 2040, emissions must be reduced by two-thirds of 80 percent below 1990 levels; and by 2050, emissions must be reduced to a level that is 80 percent below 1990 levels. The National Academy of Sciences would be the reporting agency to the EPA and Congress.

The bill also included a combination of economywide reduction targets, mandatory measures, and incentives for the development and diffusion of cleaner technologies to achieve the goals. The bill also contained the following items:

- vehicle greenhouse gas emissions standards;
- power plant greenhouse gas emissions standards;
- standards for geologic disposal of greenhouse gases;
- global warming research and development;
- energy efficiency standards in electricity generation;
- reporting system for global warming pollutants;
- clean energy task force to support development and implementation of low-carbon technology programs.

The bill was never passed into law although it was proposed in sessions of Congress for the past two years. It can be reintroduced. The measure was supported by several environmental groups, such as the Sierra Club, Greenpeace, the National Audubon Society, and the Union of Concerned Scientists.

Global Warming Wildlife Survival Act

The Global Warming Wildlife Survival Act was introduced in the House and the Senate in 2007. However, it has since died in committee.

The Consolidated Appropriations Act of 2008

The Consolidated Appropriations Act of 2008, which became Public Law 110-161 on December 26, 2007, directed the EPA to develop a mandatory reporting rule for greenhouse gases. The measure was included in a $500 billion omnibus budget that was signed into law by President Bush and will require U.S. companies to report their greenhouse gas emissions. The law did not specify, however, which industries must report or how often they must report.

Overall, the EPA would inventory approximately 85 to 90 percent of U.S. GHG emissions—from about 13,000 facilities across the nation. The GHGs included in the inventory include CO_2, methane (CH_4), nitrous oxide (N_2O), hydrofluorocarbons (HFCs), perfluorocarbons (PFCs), sulfur hexafluoride (SF_6), and other fluorinated gases, including nitrogen trifluoride (NF_3) and hydrofluorinated ethers (HFEs). Collected data will include the total GHG emissions from all sources as well as each gas by category. Once a facility has met the requirements in one year, that facility will continue to report GHG emissions in future years. Companies must reevaluate each facility's emissions whenever there is a process change or other change that may increase the facility's emissions. Facilities that fail to satisfy the reporting requirements are subject to enforcement and penalties under the Clean Air Act.

According to the EPA, data collected would be used in future policy decisions and serve as a benchmark to measure annual progress toward emissions reduction targets. This action is viewed as a first step toward a massive, comprehensive national climate change regulation.

The EPA recommends that as companies work to comply with the proposed rule, they should remain focused on the global issue of climate change and the necessity to prepare for possible further federal mandates to reduce greenhouse gas emissions. They stress that due to the importance of this issue, reducing emissions is not just a question of compliance; it is now the foundation of business performance. From now on, it should be viewed as part of the cost of doing business.

Because this act represents the first major step toward national comprehensive greenhouse gas emissions regulation, the EPA has proposed some guidelines in order to calculate an initial baseline emission measurement. Any owner or operator of a facility in the United States that directly emits GHG from specific source categories or emits 27,558 tons (25,000 metric tons) or more of CO_2 emissions annually from stationary combustion will be required to report emissions data under the regulation. The first report would be due in 2011 for calendar year 2010. Exempt from this are motor vehicle and engine manufacturers, which would start their reporting for model year 2011. The EPA has identified the following types of businesses that would be required to report their GHG emissions. (See table on opposite page.)

The effective date of this rule is 60 days after the rule is published in the *Federal Register*. The Mandatory Reporting of Greenhouse Gases Rule was published in the *Federal Register* on October 30, 2009, and became effective on December 29, 2009. The final rule was changed slightly from its April 2009 version. For example, it now exempts research and development activities from reporting, adds additional monitoring options, and requires more data to be reported rather than kept as records so that the EPA can more easily verify reported emissions.

The EPA also foresees a future role for the individual states that are already ahead in reporting and controlling emissions. It views these states as an asset for education. States could take the role in educating the

Businesses Required to Report GHG Emissions under the FY 2008 Consolidated Appropriations Act

SECTOR	REPORTERS
electricity generation	power plants
transportation	vehicle and engine manufacturers
industrial	all large industrial emitters, including those in the following industries:
• metals	iron and steel, aluminum, magnesium, ferroalloy, zinc, and lead
• minerals	cement, lime, glass, silicon carbide, pulp, and paper
• chemicals	HCFC-22, ammonia, nitric acid, adipic acid, SF6 from electrical equipment, hydrogen, petrochemicals, titanium dioxide, soda ash, phosphoric acid, electronics
• oil and gas	components of oil and gas systems (e.g., refineries), underground coal mining
other	landfills, wastewater treatment, ethanol, food processing
agriculture	manure management
upstream suppliers	petroleum refineries, gas processors, natural gas distribution companies, coal mines, importers, industrial gases

Source: Environmental Protection Agency

public and businesses and ensuring compliance. In addition, the House and Senate are currently working on a plan that is intended to position the United States as a global leader on climate change policy at the post-Kyoto discussions to take place in Copenhagen in December 2009. Progressive estimates place implementation of any such U.S. legislation dealing with climate change to take effect no later than 2012 or 2013.

American Clean Energy and Security Act of 2009

The American Clean Energy and Security Act (ACES Act, H.R. 2454) was passed by the U.S. House of Representatives by a vote of 219 to 212, on June 26, 2009. Also referred to as the Waxman-Markey Clean Energy Bill (it was proposed by Rep. Henry Waxman [D-CA] and Rep. Edward Markey [D-MA]), it contains five distinct titles: (I) clean energy, (II) energy efficiency, (III) global warming pollution reduction, (IV) transitioning to a clean energy economy, and (V) agriculture and forestry related offsets.

Title I has provisions related to federal renewable electricity and efficiency standards, carbon capture and storage technology, standards for new power plants that use coal, research and development for electric vehicles, and support for the development of the electric smart-grid. Title II provides provisions related to building, appliance, lighting, and vehicle energy efficiency programs. Title IV hosts provisions to preserve domestic competitiveness and support workers, provide assistance to consumers, and provide assistance for domestic and international adaptation initiatives. Titles III and V deal with a GHG cap-and-trade program.

The bill covers seven greenhouse gases: CO_2, methane, nitrous oxide, hydrofluorocarbons, perfluorocarbons, sulfur hexafluoride, and nitrogen trifluoride. Emitters that would be included under the regulation would include large stationary sources emitting more than 27,558 tons (25,000 metric tons) per year of GHGs; producers (i.e., refineries) and importers of all petroleum fuels; distributors of natural gas to residential, commercial and small industrial users (i.e., local gas distribution companies); producers of "F-gases"; and other specified sources. The proposal also calls for regulations to limit black carbon emissions in the United States (black carbon is formed through the incomplete combustion of fossil fuels, biofuel, and biomass and is emitted in both anthropogenic and naturally occurring soot).

The bill has set up progressive targets over time. It establishes emission caps that would reduce aggregate GHG emissions for all involved facilities to 3 percent below their 2005 levels in 2012, 17 percent below 2005 levels in 2020, 42 percent below 2005 levels in 2030, and 83 percent below 2005 levels in 2050. Commercial production and imports of

HFCs would be addressed under Title VI of the existing Clean Air Act and are covered under a separate cap.

The bill also uses the value of emission allowances to offset the cost impact to consumers and workers, to aid businesses in transitioning to clean energy technologies, to support technology development and deployment, and to support activities aimed at building communities that are more stable against climate change. It is also designed to protect consumers from higher energy prices. Low- and moderate-income households will receive a refundable tax credit or rebate. In the first few years of the cap-and-trade program, about 20 percent of the allowances will be auctioned. This percentage will increase over time to about 70 percent by 2030. The bill still needs to be voted on and passed in the Senate and signed into law by the president.

National Fuel Efficiency Policy

On May 19, 2009, President Obama—for the first time in history—set in motion a new national policy aimed at both increasing fuel economy and reducing greenhouse gas pollution for all new cars and trucks sold in the United States. The new standards, covering model years 2012 to 2016, and ultimately requiring an average fuel economy standard of 35.5 MPG in 2016, are projected to save 1.8 billion barrels of oil over the life of the program with a fuel economy gain averaging more than 5 percent per year and a reduction of approximately 900 million metric tons in greenhouse gas emissions. This would surpass the CAFE law passed by Congress in 2007 requiring an average fuel economy of 35 MPG in 2020.

"In the past, an agreement such as this would have been considered impossible," said President Obama. "That is why this announcement is so important, for it represents not only a change in policy in Washington, but the harbinger of a change in the way business is done in Washington. And at a time of historic crises in our auto industry, this rule provides the clear certainty that will allow these companies to plan for a future in which they are building the cars of the 21st century."

President Obama also said that the changes necessary to achieve better efficiency would cost consumers an extra $1,300 per vehicle starting in 2016, but drivers would be saving at the pump. He estimated that a typical driver would save $2,800 over the lifetime of a car, assuming gasoline costs around $3.50 per gallon by then. He also stressed that the increased miles per gallon should cut greenhouse gas emissions by more than 992 million tons (900 million metric tons), which is equivalent to shutting down 194 coal plants. What the plan means for mileage per gallon is as follows: while the 30 percent increase translates to a 35.5 MPG average for both cars and light trucks, the percentage increase in cars would be greater, rising from the current 27.5 MPG standard to 39 MPG starting in 2016. The average for light trucks would rise from 24 MPG to 30 MPG. For 2009 car models, however, according to MSNBC (5/19/09), the industry has really averaged 32.6 MPG; and if all goes as planned, by 2016 Americans can expect dozens of hybrid, plug-in hybrid and even all-electric vehicle models. The national program will be finalized once the Department of Transportation and the EPA finalize the specifics, followed by a public review period.

Cap and Trade and Other Mitigation Strategies

Throughout the United States and the world, regions are adopting policies in an attempt to make progress against climate change. Positive actions include increasing renewable energy generation, selling agricultural carbon credits, and encouraging energy efficiency. The positive effects of these are reducing vulnerability to energy price spikes, promoting development of local economies, and improving air quality. This chapter examines cap and trade as a policy tool and how the carbon trading market works in an international arena and looks at the need for global action and what will be the economic implications. It also explores some of the activities individual states are undergoing in an effort to combat global warming.

CAP AND TRADE

Cap and trade is "an environmental policy tool that delivers results with a mandatory cap on emissions." The cap is the foundation on which the policy is constructed—it is the permissible carbon emission limit. In other words, for a country, it is the absolute, nationwide limit on global warming pollution. This measurement is usually set on a scale of billions of tons of CO_2 (or equivalent for other greenhouse gases [GHGs]) released into the atmosphere each year. Once the cap is in place and is being met, then over time, the cap is lowered in order to further cut emissions; the principal objective being to lower it enough over time to avoid the worst consequences of global warming.

The trade portion of the system is a market created by powerful incentives for companies to reduce the pollution they would normally emit. The trade market also works with the individual emitters and provides flexibility in how they can meet their limits. In order to make all this happen within a country, such as the United States (each country under the Kyoto Protocol has a specific emission reduction level they are working toward), the respective government creates allowances that add up to the total emissions allowed under the cap. Each year, those industries and businesses subject to the cap must turn in allowances equal to their emissions for that year. Examples of industries and businesses that must do this include power plants, manufacturing industries, chemical industries, steel industries, mining companies, processing industries, and any other entity that produces and releases large amounts of CO_2 into the atmosphere. In order for the nation to meet the cap, each of these entities must reduce their emissions. If an entity reduces its emissions enough that it has more allowances than it needs, it can profit by selling the extra allowances. This opportunity gives them the incentive to reduce their emissions below what is mandated by the cap.

If an entity finds it too expensive to reduce its emissions, cap and trade allows it to purchase more allowances from other entities that have reduced their emissions far enough that they have extra allowances. The more a company reduces its emissions the more money it can either make or save.

Cap and trade works internationally the same as when applied within a single country. Under the Kyoto Protocol, countries required to reduce their emissions are allowed to purchase carbon credits from developing countries or from industrialized countries whose emissions are below the level required. The credits cover emissions of all GHGs, which are expressed as carbon dioxide equivalents (CO_2e).

According to the Environmental Defense Fund (EDF), credits applying to any GHG are a serious limitation to the policy, and they believe they should only be used for specific types of pollution. The EDF says that CO_2 travels quickly to the upper atmosphere and does not become concentrated in one particular area of the landscape. Emissions such as mercury, however, are usually deposited near where they are emitted, creating hotspots. Because mercury is also a toxin that poses a threat to human health, it should not be included in cap and trade.

The international trade in carbon credits is intended to promote investment in energy efficiency, renewable energy, and other ways of reducing emissions. In the majority of developed, industrialized countries, GHG-emitting companies have taken on the responsibility of running, regulating, and facilitating the trade of carbon credits in the carbon market. There are two main types of carbon markets: (1) project-based markets, and (2) allowance-based markets.

Project-based Markets

Project-based markets encourage investment in companies or programs that are committed to reducing emissions. These projects are run under the clean development mechanism (CDM) or joint implementation (JI). The CDM is an arrangement under the Kyoto Protocol that allows industrialized countries with a GHG-reduction requirement (called an Annex B party) to invest in projects that reduce emissions in developing countries as an alternative to more expensive emission reductions in their own countries. The critical factor that distinguishes an approved CDM carbon project is that it must prove its actions have reduced emissions in the developing country that would not have occurred otherwise; this is a concept called additionality. What the CDM does in effect is allow net global GHG emissions to be reduced at a much lower global cost through the financing of emissions reductions projects in developing countries where the costs are much lower than they would be in industrialized countries. The CDM is supervised by the CDM executive board (CDM EB) and is overseen by the Conference of the Parties (COP) of the United Nations Framework Convention on Climate Change (UNFCCC). According to the UNFCCC, the CDM is viewed as a trailblazer. It is the first global environmental investment and credit scheme of its kind, providing a standardized emissions offset plan. An example of a CDM might involve, for example, a rural electrification project using solar panels or the installation of more energy-efficient boilers. The UNFCCC views CDM as a way to stimulate sustainable development and emission reductions, while giving industrialized countries some flexibility in how they meet their emission-reduction or limitation targets.

The JI in the Kyoto Protocol helps countries (Annex I countries, see chapter 1) with GHG caps to meet their obligations. Any Annex I country can invest in emission reduction projects (called joint implementation projects) in any other Annex I country as an alternative to reducing emissions domestically. This mechanism allows countries to lower the costs of complying with their respective Kyoto targets by investing in GHG reductions in an Annex I country where reductions are cheaper and then applying the credit for those reductions toward their commitment goal. An example of a JI project could involve replacing a coal-fired power plant with a more efficient combined heat and power plant or a coal-heated building with a geothermal-heated building. JI projects are undertaken in countries that have economies in transition. JI projects differ from CDM projects in that JI projects are done in countries that have an emission-reduction requirement.

Through a JI project, emission reductions are awarded credits called emission reduction units (ERUs), where one ERU represents an emission reduction equaling 1.1 tons (1 metric ton) of CO_2e. The ERUs come from the host country's pool of assigned emissions credits, known as assigned amount units (AAUs). Each Annex I party has a predetermined amount of AAUs that are calculated on the basis of its 1990 GHG emission levels. By requiring JI credits to come from a host country's pool of AAUs, the Kyoto Protocol ensures that the total amount of emissions credits among Annex I parties does not change for the duration of the Kyoto Protocol's first commitment period.

JI offers a flexible, cost-effective means of fulfilling part of their Kyoto commitments, while the host country (receiver) benefits from both foreign investment and technology transfer. A JI project must provide a reduction in emissions by sources or an enhancement of removal by *sinks* that is additional to what would otherwise have occurred.

As far as project-based markets using CDM or JI mechanisms, the main buyers today are the industrialized and transition economies; the principal sellers are in Asia and South America, with India and Brazil in the foreground.

Allowance-based Markets

Allowance-based markets are what enable large companies—such as energy producers—to purchase emission allowances under plans administered by international carbon trading organizations, such as the EU Emissions Trading System (EU ETS). Allowance-based markets enable companies to offset their emissions by purchasing credits from countries that either have no limit placed on their emissions or have kept emissions below the level required. Since 2003, partly spurred on by the EU Emissions Trading Scheme's opening in 2005, the carbon trading business has been growing. Carbon trading schemes are now opening up worldwide and include the Carbon TradeEx America and the Chicago Climate Exchange (CCX), which was established by several large corporations along with the World Resources Institute.

The concept of carbon markets is still fairly new. Today, they only account for roughly 0.5 percent of the annual global GHG emissions. The idea is gaining in popularity and is being recognized as an effective global tool for slowing global warming. Especially encouraging is the fact that carbon trade is now being conducted within the United States, which is not a participant in the Kyoto Protocol.

Carbon credits are sold in 100-ton (91 metric-ton) units. If a business is selling credits but does not have 100 tons, then the carbon trading company combines more than one available partial unit together to make a salable unit. There is still debate on what is tradable and how concrete an emissions reduction a given practice achieves. To deal with uncertainty, some practices are discounted. On a farm, for example, tradable units considered include the following:

- capture of methane from a waste lagoon/anaerobic digester;
- practice of no-till to sequester carbon on large acreages;
- reduction of *nitrogen* application to reduce nitrous oxide emissions and energy;
- practice of timber stand improvement in woodlands to sequester carbon in trees;

- supply of an energy processor with wood chips, grass for pellets, oil seed for biodiesel, etc., to displace fossil fuels;
- completion of improvements in efficiency, reducing energy use;
- use of wind, solar, or geothermal energy sources to displace fossil fuel use.

While carbon trading is a futures market, the rules of the game are still being developed. Income generated from carbon trading could help pay for adoption of new practices and keep farms or land financially viable.

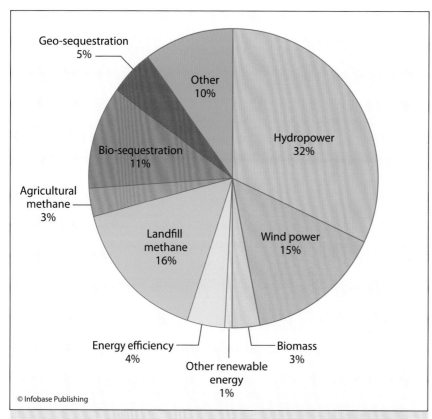

Trading carbon credits is one way to share the burden of reducing CO_2 emissions globally. The types of projects where carbon credits were traded in 2004–2005 are shown in the graph.

The Economics of Cap and Trade

According to Nat Keohane, Ph.D., director of economic policy and analysis at EDF, aggressive cap and trade is not only affordable, but critical to both the Earth and humanity's future. The cost to the economy will be minimal—it will be less than 1 percent of the U.S. gross domestic product (GDP) in 2030. Keohane also stresses that the longer action is delayed the more expensive it will be to make emission cuts. In addition, the more time that passes without addressing the issues, the more irreversible damage will be done by global warming.

Through the use of economic models, Dr. Keohane determined that by continuing with a business as usual approach, the U.S. economy would reach $26 trillion by January 2030. With a cap on GHG emissions, however, the economy will reach the same level only two to seven months later. Therefore, the impact on the economy would not be that significant—"just pennies a day," according to Dr. Keohane.

He also stresses that total job loss would be minimal (the manufacturing sector would experience some impact), and the new carbon market would create a multitude of new jobs. He said that American households will be most affected by energy costs, but even there the increase would be modest. Overall costs would be small enough to allow programs to be developed that would take any burden off low-income households.

Dr. Keohane believes that cap and trade is the best means to fight global warming because it not only gives each company the ability to choose how to cut their emissions, it gives the economy the most flexibility to reduce pollution in the most cost-effective way. He also says it turns market failure into market success: "Global warming is a classic example of what economists term 'market failure.'" GHG emissions have skyrocketed because their hidden costs are not factored into business decisions—factories and power plants pay for fuel but not for the pollution they cause. Putting a dollar value on the pollution fixes that failure and gives industry incentive to pollute less.

"It also taps American ingenuity. History shows that Americans can overcome steep challenges. In two short years during World War II, Americans redirected much of the U.S. economy. Manufacturers produced different goods against tight deadlines. Detroit converted car factories to munitions production. Fireworks factories made military explosives. A. C. Gilbert, a maker of model train engines, produced airborne navigational instruments. Given the right incentives, we can transform the way we make energy too.

"But we must act immediately, or costs and risks will rise. Costs will remain low only if we act quickly. The longer we wait to curb pollution, the steeper the cuts must be to avoid catastrophic climate change. We need time to develop new technologies and build infrastructure. Plus, developing countries like China and India are waiting for us to act before they take action. We have very little time remaining to cap greenhouse gas emissions before we incur a large risk of climate catastrophe, heavy economic costs, or both. But if we start now, we can do it—affordably."

STATE MITIGATION PROJECTS

While Congress has lagged in effort to tackle the global warming issue through legislation, several individual states have stepped up and taken a leading role in combating the issue within their jurisdictions. Each region has its own GHG emission profile to deal with, as different sectors of the economy emit different GHGs, making each state's response plan unique. This section highlights some of the states' accomplishments.

California

California's governor Arnold Schwarzenegger has taken significant steps to confront the global warming issue and proven to be one of the leaders in the United States in taking action. In San Francisco on June 1, 2005, he announced his Environmental Action Plan that seeks a reduction of California's (the most populated U.S. state) GHG emissions 80 percent below 1990 levels by 2050. Taking likely population growth into account, this may require a cut in per capita emissions of more than 90 percent in some areas. A three-tiered plan, it was announced the day before the opening of a UN Conference on Green Cities hosted by San Francisco mayor Gavin Newsom.

Schwarzenegger signed an executive order setting out his environmental action plan. In Phase I, it seeks to reduce California's GHGs by 2010 to less than 2000 levels, and in Phase II to reduce emissions by 2020 to less than levels in 1990. Taking into account both the population and

Cap and Trade and Other Mitigation Strategies

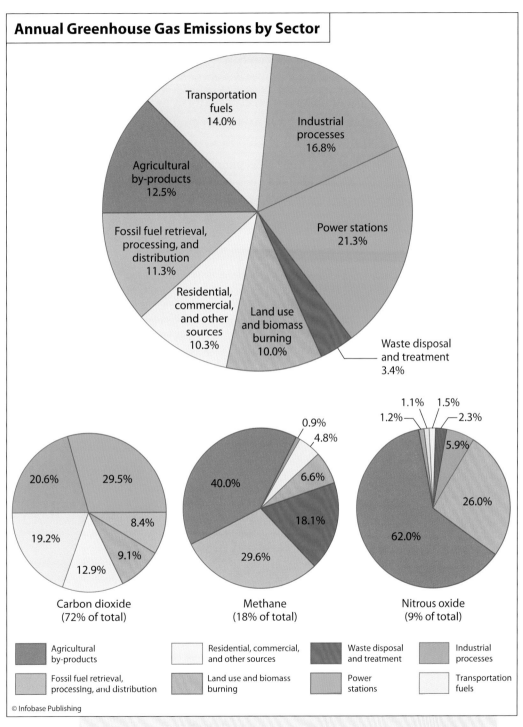

Greenhouse gas emissions by sector—the smaller pie charts break down the GHGs in specific gases by sector

economic growth expected to occur in California, politicians and environmentalist realize how critical it is to make much more efficient use of their processes by immediately shifting to less carbon-intensive fuels. Phase III aims to achieve an 80 percent reduction below 1990 levels by 2050.

This is by far the most ambitious plan ever set forth by an executive branch political leader in any major industrialized nation. Schwarzenegger says the goals will be met through existing and future technology, as well as government-backed incentives. His plan also sets these goals:

1. cut air pollution statewide by up to 50 percent and restore independence from foreign oil
2. invest in hydrogen highways
3. fight for federal dollars for hydrogen fuel development
4. expedite clean fuel transportation
5. remove gross-polluting vehicles from the road
6. protect California's air-quality standards from industrial facilities
7. relieve traffic congestion
8. protect California's rivers, bays, and coastline
9. reduce ocean pollution
10. protect drinking water
11. solve California's electrical energy crisis
12. promote solar renewables
13. increase the reliability of the grid
14. save energy through green building
15. increase renewable energy
16. improve mass transit

At the UN Conference on Green Cities, San Francisco mayor Gavin Newsom said, "The reality is, in cities we consume some 75 percent of the world's natural resources. And as a consequence, and by extension, we pollute disproportionately the world as it relates to the consumption of those resources. But the good news is, as mayors around the world know all too well, and the former mayors know, you can do an extraordinary amount without waiting around for someone else to solve the problem, at the local level."

Previous California governor Gray Davis also made an admirable attempt to solve the global warming issue. He signed a landmark bill—called the California Climate Bill—into law on July 22, 2002, designed to cut car exhaust emissions. This regulation required automakers to limit emissions of GHGs in an effort to curb global warming. This law was possible in California because that state has a unique loophole that allows them to set their own air-quality standards independent of the federal government. The law requires the California Air Resources Board (CARB) to obtain the "maximum feasible" cuts in GHGs emitted by all noncommercial vehicles (including cars, light-duty trucks, and sport utility vehicles) in model year 2009 and beyond. The standards apply to automakers' fleet averages, rather than each individual vehicle, and carmakers will be able to partially achieve the standards by reducing pollution from nonauto sources, such as factories. The regulations officially took effect on January 1, 2006, but gave automakers until 2009 to come up with technological changes or modifications to comply with the new standards.

According to Davis, "This is the first law in America to substantively address the greatest environmental challenge of the 21st century. In time, every state—and hopefully every country—will act to protect future generations from the threat of global warming. For California, that time is now."

In passing this bill, California was the first state to require catalytic converters, unleaded gasoline, and smog checks.

Gray Davis signed another landmark bill requiring that a minimum of 20 percent of California's energy come from renewable resources. SB 1078, which sets the California renewables portfolio standards, requires retail sellers of electricity to produce 20 percent of their electricity from renewable resources by 2017. Sellers must increase their use of renewable energy sources by no less than 1 percent per year moving toward 20 percent.

Because of this bill, California has shown itself to be both a national and world leader in reducing greenhouse gases by passing groundbreaking laws that establish the highest renewable energy requirement in the nation. To subsidize the program, a fee that utility consumers are already paying will finance it. Under the bill, however, utility companies cannot use hydropower to meet the new goal because of concern about the impact of hydropower on the environment; the renewable energy must be from solar, wind, geothermal, and other renewable sources.

Washington

In February 2007, Washington State's governor Chris Gregoire and four other western governors committed to join forces to reduce GHG emissions. According to an article in the *Seattle Post-Intelligencer* on February 27, the governors of Washington, Oregon, Arizona, California, and New Mexico formed the Western Climate Initiative (WCI). Their goals are to:

1. create a regional target for reducing GHG emissions within six months
2. establish the means for meeting these goals over the next 18 months (possibly through cap and trade)
3. create a registry for tracking and managing GHG emissions

Their plans include cutting GHG emissions to 50 percent below 1990 levels by 2050. Washington has also adopted the tough emission standards that Schwarzenegger put into effect in California and approved emission caps that will come into effect in 2012 for some pollution sources. Senator Erik Poulsen (D-WA) said, "The real work that must happen in this arena is to have fewer and cleaner cars. Until we get more serious about public transportation, we're only going to make a dent in the problem."

The WCI's Electricity Committee and partners are currently working on their regional cap-and-trade program to decide on issues such as point of regulation for imported electricity, compliance enforcement options, and practical and administrative aspects. They have also just released the final version of the first group of Essential Requirements for Mandatory Reporting (ERMR). This release includes topics such as: general provisions governing all reports; requirements for third-party verification; greenhouse gas monitoring; and reporting and record-keeping methodologies for various source categories, such as: fuel combustion, electricity generation, aluminum manufacturing, cement manufacturing, coal storage, iron and steel manufacturing, lime manufacturing, petroleum refining, pulp and paper manufacturing, soda ash production, and petrochemical production. Rules implementing these essential requirements for the 2010 reporting year will be put in place shortly.

The Western Governors' Association

The Western Governors' Association (WGA) is a coalition of governors from 19 states in the western United States (Alaska, Arizona, California, Colorado, Hawaii, Idaho, Kansas, Montana, Nebraska, Nevada, New Mexico, North Dakota, Oklahoma, Oregon, South Dakota, Texas, Utah, Washington, and Wyoming) and three U.S.-flag Pacific Islands (American Samoa, Guam, and Northern Mariana). Its mission is to address important policy and governance issues in the West, to strengthen the economy of the region, and to develop policy and carry out programs in the areas of natural resources and the environment. It also strives to be a source of innovation and promote development of solutions to regional problems.

The WGA is currently involved with the global warming issue. It has called for a national energy plan that will provide affordable and clean energy able to sustain the economy, stimulate greater energy efficiency, strengthen energy security and independence, and reduce GHG emissions. They are currently working with the U.S. Department of Energy (DOE) on the Western Renewable Energy Zones (WREZ) initiative, which is geared to speed up the development and delivery of electricity generated by renewable energy throughout the western United States.

The WGA has also put in place the Clean and Diversified Energy Initiative (CDEi), with the goal of addressing climate change issues on three fronts. First, it promotes the widespread adoption of energy-efficiency measures. It also promotes the use of renewable energy, such as wind, solar, geothermal, and biomass. Third, it looks at coal-generating plants and the technology of carbon capture and *sequestration* as a possibility for the future. It also formed the Western Regional Air Partnership (WRAP), which is an agreement between federal agencies, state governments, and tribal governments to develop the technological and policy tools necessary to keep air quality at healthy levels.

Currently, WRAP has completed two climate change processes: the Greenhouse Gas Emissions Inventories and Forecasts and the Climate Registry (a greenhouse gas registry involving multiple states, Canadian provinces, and the Mexican state of Sonora). Other projects they have been involved in include transportation fuels for the future and future drought management in the West.

ECONOMICS OF MITIGATION

According to the International Monetary Fund (IMF), it is possible to fight global warming without having a negative impact on economic growth. While the IMF reports that "Climate change is a potentially catastrophic global externality and one of the world's greatest collective action problems," in order to curb global warming the IMF suggests a worldwide long-term plan of gradual increases in carbon prices. If this happened, they believe it would bring about the needed shifts in investments and consumption; it would discourage people from buying emission-intensive and energy-inefficient products. For example, if there is a better financial incentive to purchase a fuel-efficient hybrid car over a gas guzzler, people will buy the hybrid. The IMF has also determined that mitigation would not have as drastic an impact on the world economy as some fear.

The *Stern Review on the Economics of Climate Change* is one of the best-known reports on the economics of global warming. The 2006 report was prepared by Nicholas Stern, Baron Stern of Brentford, former head of the UK Government Economic Service and former World Bank Chief Economist, now I.G. Patel Professor at the London School of Economics and Political Science. The *Stern Review* predicts that climate change will have a serious impact on economic growth without mitigation and recommends that 1 percent of the global GDP be invested to mitigate its effects. If this is not done, it could cause a recession equivalent to upwards of 20 percent of the GDP.

The insurance industry is very concerned about the economic implications of global warming. Since 1960, the number of major natural disasters has tripled. Over the past 30 years, the proportion of the global population affected by weather-related disasters has doubled, rising from 2 percent in 1975 to 4 percent in 2001.

A 2005 report from the Association of British Insurers concluded that limiting carbon emissions could avoid up to 80 percent of the projected additional annual costs of tropical cyclones by the 2080s. In June 2004, a report issued by the Association of British Insurers stated, "Climate change is not a remote issue for future generations to deal with. It is, in various forms, here already; impacting on insurers' businesses now."

The world's two largest insurance companies—Munich Re and Swiss Re—stated in a study released in 2002 that "The increasing frequency of severe climatic events, coupled with social trends, could cost almost $150 billion each year in the next decade. These costs would, through increased costs related to insurance and disaster relief, burden customers, taxpayers, and industry alike."

The costs to mitigate global warming depend on several factors. The most important factor is the target level of CO_2. The lower the level (in ppm), the sooner steps must be taken to reach that goal. The sooner action must be taken and results achieved, the shorter the interval over which the costs must be spread, which makes initial mitigation more expensive. A commonly referenced target level by many countries is 550 ppm (current levels are 380 ppm), but the level is rising an average of 2–3 ppm annually. Nations that signed the Kyoto Protocol are required to lower their emissions to a specific level below their 1990 emissions level.

In terms of potential mitigation costs, the Intergovernmental Panel on Climate Change (IPCC) has estimated annual mitigation costs could range from $78 billion to $1,141 billion, roughly equal to 0.2–3.5 percent of the current world GDP (which is approximately $35 trillion).

In 2008, the McKinsey Global Institute used a cost curve analysis to determine that it was possible to stabilize global GHG concentrations at 450 to 500 ppm with costs around 0.6–1.4 percent of the global GDP by 2030.

In 2007, the chairman of Lloyd's of London, Lord Peter Levene, stated, "The threat of climate change must be an integral part of every company's risk analysis."

The former U.S. vice president Al Gore recently challenged the nation to produce every kilowatt of electricity through wind, solar, and other renewable sources of energy. He believes the cost of switching to clean electricity sources could cost as much as $3 trillion over 30 years in public and private money. Even though this seems like a lot, he also says that the investment will "pay itself back many times over," and that he has "never seen an opportunity for the country like the one that is emerging now."

Another suggestion presented as a way to finance mitigation is to create a new global warming tax. A *New York Times* article from September 16, 2007, outlines how Gilbert Metcalf, an economics professor at Tufts University, describes how revenue from a carbon tax could be used to actually reduce payroll taxes in a way that would leave the distribution of the total tax burden basically unchanged. He proposed a tax of $15 per metric ton of CO_2, together with a rebate of the federal payroll tax on the first $3,660 of earnings for each worker. Proponents of this scheme feel this is a better approach than forcing cars to become more fuel-efficient. Their argument is that when cars get better mileage, the owners will just be tempted to drive them more often, rather than cut back and start using public transportation.

Another issue concerning the economics of mitigation is the state of the present U.S. economy. Namely, the current recession may very well affect funding for poor nations to fight global warming. Currently, African activists have appealed for major polluters—like the United States—to commit to donating 1 percent of their GDP toward foreign mitigation efforts.

The United Nations Climate Change Conference was held in Copenhagen, Denmark, on December 7–18, 2009. Attending the conference were the leaders of 192 countries. World leaders saw this internatonal gathering as a critical step forward toward the solution to global warming. It provided a framework to determine how best to deal with climate change after the Kyoto Protocol expires in 2012. It is essential that this be accomplished with a global perspective, enabling all countries to willingly participate in long-term workable solutions. This conference provided hope and a yardstick for establishing concrete and measurable actions and goals.

Antonio Hill, a senior policy adviser for the British aid group Oxfam, expressed concern with the situation, especially the fact that wealthy nations were willing to lend developing nations money but were less willing to donate. Hill remarked, "As far as we're concerned this is the moral equivalent of having someone drive a car into your house and offering you a loan to pay for the damages."

If a multilateral policy is going to work, all countries must participate because emerging and developing economies are expected to produce 70 percent of global emissions during the next 50 years. In addition, any framework that does not include large and fast-growing economies (China, India, Brazil, and Russia) would be very costly and politically unwise.

This chapter has presented several financial and technological strategies to handle the mitigation of global warming. Whichever methods are used will ultimately depend on the region, available technology, available finances, and political policy. What is critical is that action be taken immediately to fight climate change in order to lower the negative consequences of sea-level rise, flooding, drought, disease, and other disasters.

The International Political Arena

Because global warming is a global problem, it will take a global solution. It does not matter whether greenhouse gases are released in Los Angeles, London, Tokyo, or Paris, they have the same impact on the atmosphere. Thus, if only a few countries make an effort to slow emissions, it will not solve the global warming problem. All countries must be involved in the solution in order to successfully solve the problem. This chapter looks first at the opinion of one of the world's leading experts on global warming concerning the ramifications of holding off on taking action. Next, it presents an overview of how international cooperation eventually evolved and the events that fueled it. The chapter then examines the unique role of international organizations and what they have accomplished and finally focuses on the progress of individual countries and regions.

AN EXPERT'S WARNING

Dr. James E. Hansen, one of the world's foremost experts on global warming, cautions that the world has only a 10-year window of opportunity left to take decisive action on global warming and still avoid catastrophe. Hansen, the longtime head of the NASA Goddard Institute for Space Studies (GISS) tells governments that they must put plans in place now in order to keep CO_2 emissions under control so that temperatures do not increase any more than 1.8°F (1°C).

In attendance at the annual Climate Change Research Conference in September 2006, Dr. Hansen said, "I think we have a very brief window of opportunity to deal with climate change . . . no longer than a decade at the most. If the world continues with a business as usual scenario, temperatures will rise by 3.6–7.2°F (2–3°C) and we will be producing a different planet." Changes he noted include the rapid melting of ice sheets, rising sea levels that would flood areas like Manhattan, prolonged droughts, deadly heat waves, powerful hurricanes in places they had never occurred before, and the likely extinction of 50 percent of the world's species.

Two major actions Hansen advocates are to increase energy efficiency and reduce dependence on fossil fuels. Hansen focused on the Arctic ecosystem because it was one of the first areas to show the effects of global warming. "It is not too late to save the Arctic, but it requires that we begin to slow carbon dioxide emissions this decade."

Mark Serreze, a senior research scientist at the National Snow and Ice Data Center in Boulder, Colorado, says, "The latest findings are coming in line with what we expected to find. We're starting to see a much more coherent and firm picture occurring."

Loss of summer sea ice means less sunlight gets reflected, lowering the Arctic's albedo, and more gets absorbed, adding to the global warming problem. Besides melting sea ice, it threatens Arctic wildlife. In fact, the polar bear population in Canada's Hudson Bay has taken an especially hard hit. Dr. Nick Lunn of the Canadian Wildlife Service (CWS) determined that the polar bear population in the Western Hudson Bay region has declined 22 percent in the past 17 years, from 1,200 to less than 1,000. A report issued by the USGS in 2009 (Polar Bear Population Status in Southern Hudson Bay, Canada) voices the same conclusion. In addition, the CWS has collected overwhelming evidence that the condition of adult bears has been steadily decreasing, with the average

The polar regions are being hit the hardest by global warming. If corrective action is not taken this decade, the polar bear could become extinct. *(Fotosearch)*

weight of females declining toward a threshold at which the chances of it being able to bear viable cubs is becoming doubtful. Dr. Lunn has concluded that the threshold may be reached, if the trends continue as they have, as soon as 2012.

The primary cause for the deteriorating condition of this population of bears is the early breakup of Arctic sea ice. Bears have to go farther and work harder to find their principal source of food—the ring seal. Because of this, when females give birth, they are much more emaciated than normal and have a more difficult time feeding their cubs and giving them proper nutrition. As a result, more cubs are not surviving to adulthood. The overall threat to the population is that the current generation of bears will not be replaced.

"The Western Hudson Bay region is one of the most studied populations in the world, so the data set for these bears is the most complete and accurate available. The low Arctic region they inhabit is an ecosystem highly vulnerable to climate change, and so it is likely that what we are seeing with this population will continue to spread throughout all circumpolar bear populations as environmental changes in the north accelerate. The polar bear, is, of course, just one aspect of a finely balanced and fragile ecosystem; one that is stressed and changing fast. We ignore those changes in the Arctic, to the polar bear, and all that supports and depends on it, to our own peril. As goes the polar bear, we have to wonder, goes the rest of the world?" Dr. Lunn said.

THE EVOLUTION OF INTERNATIONAL COOPERATION

The space exploration era not only gave scientists a new view of the Earth and global science, but data began to be recorded in new ways. Computers and modeling software led to new studies and discoveries, some of the most interesting findings were the changing levels of CO_2 in the atmosphere (Keeling's curve in 1958), climate cycles, paleoclimatology through interpretation of ice cores, and ocean/atmospheric circulation patterns.

In the late 1960s, an environmental movement was gaining momentum worldwide, and climate change became one of the most-discussed topics. The first significant conference where scientists discussed climate change was the Global Effects of Environmental Pollution Symposium held in Dallas, Texas, in 1968. Then, in 1970, a monthlong Study of Critical Environmental Problems (SCEP) at the Massachusetts Institute of Technology (MIT) was held. At this symposium, nearly all of the attendees were from the United States, and they felt the need for better international representation. This led to a second gathering in which 14 nations met in Stockholm in 1971, where they discussed climate change—a Study of Man's Impact on Climate (SMIC).

Each attendee returned home with a dire message to their nation: Rapidly melting ice and rapid climate change could occur in the next

100 years because of human activity. The recommendation of the scientists was to create a major international program to monitor the environment. From this recommendation, the United Nations Environment Programme (UNEP) was formed.

At this point, researching climate and gathering data had officially become one of the UN's environmental responsibilities. One of the milestones at the time was that the scientists involved pointed out that "the rate and degree of future warming could be profoundly affected by government policies." They called on governments to consider positive actions to prevent future warming. This was the tipping point where climate science shifted from a merely scientific issue to a political issue. As a result, in 1986, a small committee of experts, the Advisory Group on Greenhouse Gases (AGGG), was formed.

This spurred international, national, and regional conferences, which further promoted research and scientific collaboration. The result in the 1980s was interesting. Studies, research, and conferences conducted by organizations such as the U.S. National Academy of Sciences gained momentum among climate scientists. According to the science writer Jonathan Weiner, "By the second half of the 1980s, many experts were frantic to persuade the world of what was about to happen. Yet they could not afford to sound frantic, or they would lose credibility."

One of their big fears was that any push for policy changes would set the scientists against potent economic and political forces and also against some colleagues who vehemently denied the likelihood of global warming. The scientific arguments became entangled with emotions.

What was called for was more proof—more concrete data. So the scientists went back to work. New research concepts were developed. Scientists began looking at the issue as a climate system, using the input of all related scientific fields (geophysics, chemistry, biology, etc.). By looking at everything together, computer models could be developed to begin understanding how global warming worked and therefore how it could be prevented.

In 1982, through scientific work conducted by the UNEP, the Vienna Convention for the Protection of the Ozone Layer was held, and 20 nations signed the document created at the convention. When the ozone hole was discovered over Antarctica and shocked the world, it led to the 1987 Montreal Protocol of the Vienna Convention, where governments formally pledged to restrict emissions of specific ozone-damaging chemicals. The Montreal Protocol has had great success in reducing emissions of Chlorofluorocarbons (CFCs) and further damaging the ozone layer. It has not, however, had a significant impact toward reducing global warming.

The success at Montreal was followed up the next year by a World Conference on the Changing Atmosphere: Implications for Global Security, also called the Toronto Conference. The conclusions drawn at this conference were that "the changes in the atmosphere due to human pollution represent a major threat to international security and are already having harmful consequences over many parts of the globe."

For the first time, a group of prestigious scientists called on the world's governments to set strict, specific targets for reducing greenhouse gas (GHG) emissions. They advised that by 2005, the world should push its emissions 20 percent below the 1988 level. Observers saw this goal as a major accomplishment, if only as a marker to judge how governments responded.

The Toronto Conference caught many politicians' attention. Officials were impressed by the warnings of prestigious climate experts. Prime Minister Margaret Thatcher, herself a chemist, gave global warming official endorsement when she described it as "a key issue" in a speech she delivered to the Royal Society in September 1988. At that time, she also increased funding for climate research. She was the first major world leader to take a positive, strong position to do something to fight global warming.

In 1988, the World Meteorological Organization (WMO) and the UNEP collaborated in creating the Intergovernmental Panel on Climate Change (IPCC). Unlike earlier conferences, national academic panels, and advisory committees, the IPCC was composed mainly of people who participated not only as science experts, but as official representatives of their governments—people who had strong links to national offices, laboratories, meteorological offices, and scientific research agencies like NASA. Today, most of the world's climate scientists are involved in the IPCC, and it has become a pivotal player in policy debates. Since 1988, global warming has been accepted as an international issue, both scientifically and politically.

THE ROLE OF INTERNATIONAL ORGANIZATIONS

An evolution of events led to the productive international cooperation that could effectively deal with global warming. Once international cooperation had been put in place, the creation of international organizations naturally followed. This section discusses some of those organizations.

Renewable Energy and Energy Efficiency Partnership

The Renewable Energy and Energy Efficiency Partnership (REEEP) is a worldwide public-private partnership that was originated by the United Kingdom, other business interests, and governments at the Johannesburg World Summit on Sustainable Development (WSSD) in August 2002. Its goals are to reduce GHG emissions, help developing countries by improving their access to reliable, clean energy, make renewable energy and energy efficiency systems (REES) more affordable, and help nations financially who engage in energy efficiency and use renewable resources.

The United Kingdom's rational for developing REEEP was an effort to correct the fact that there was nothing else in place—either policy-wise or regulatory—to promote renewable energy or energy efficiency. In addition, it was felt that current limits in a country's finances stood in the way of being able to make the transition, and economic assistance was needed. By removing these market barriers, it was hoped that more progress would be made toward achieving the long-term transformation of the energy sector.

REEEP relies on a bottom-up approach, where partners work together at regional, national, and then international levels to create policy, regulatory, and financing programs to promote energy efficiency. Currently, REEEP is funded by many governments, including Australia, Austria, Canada, Germany, Ireland, Italy, Spain, the Netherlands, the United Kingdom, the United States, and the European Commission. The European Commission is the executive branch of the European Union of which 27 countries are members (Austria, Belgium, Bulgaria,

Cyprus, Czech Republic, Denmark, Estonia, Finland, France, Germany, Greece, Hungary, Ireland, Italy, Latvia, Lithuania, Luxembourg, Malta, Netherlands, Poland, Portugal, Romania, Slovakia, Slovenia, Spain, Sweden, and the United Kingdom).

REEEP currently has nearly 50 ongoing projects covering roughly 40 countries including China, India, Brazil, and South Africa. They work with 202 partners, 34 of whom are governments (including all the G8 countries, except Russia), countries from emerging markets and the developing world, businesses, nongovernmental organizations (NGOs), and civilian volunteers. REEEP relies on partners' voluntary financial contributions, experience, and knowledge.

European Climate Change Programme

The European Climate Change Programme (ECCP) was begun in June 2000 by the European Union's European Commission. Their goal was to identify, develop, and implement all the necessary elements of an EU strategy to implement the Kyoto Protocol. All EU countries' ratifications of the Kyoto Protocol were deposited on May 31, 2002.

The EU decided to work as a unit to meet its Kyoto emissions targets. The ECCP approaches this by using an emissions scheme known as the European Union Emissions Trading System (EU ETS). In order to achieve their legally binding commitments under Kyoto, countries have the option of either making these savings within their own country or buying emissions reductions from other countries. The other countries still need to meet their Kyoto target reductions, but the use of a free market system enables the reductions to be made for the least possible cost. Most reductions are made where they can be made in the least expensive manner, and excess reductions can be sold to other countries whose cuts are prohibitively expensive.

EU ETS is the largest GHG emissions trading scheme in the world. In 1996, the EU identified as their target a maximum of 3.3°F (2°C) rise in average global temperature. In order to achieve this, on February 7, 2007, the EU announced their plans for new legislation that required the average CO_2 emissions of vehicles produced in 2012 to exceed no more than 130 g/km. Looking ahead to the time when the

Kyoto Protocol expires in 2012, the ECCP has identified the need to review their progress and begin creating a plan of action to implement once the Kyoto Protocol expires. To launch their "post–2012 climate policy" the EU held a conference on October 24, 2005, in Brussels. From this, the Second European Climate Change Programme was launched. The ECCP II consists of several working groups:

1. the ECCP I review group (comprised of five subgroups: transport, energy supply, energy demand, non–CO_2 gases, and agriculture)
2. aviation
3. CO_2 and cars
4. *Carbon capture and storage* (CCS) technology
5. adaptation
6. EU emissions trading schemes

Some of the highlights of their work follow. In their assessment of aviation, the EU determined that it contributes to global climate change and its contribution is increasing. Even though the EU's total GHG emissions fell by 3 percent from 1990 to 2002, emissions from international aviation increased nearly 70 percent. Even though there have been significant improvements in aircraft technology and operational efficiency, it has not been enough to neutralize the overall effect of aviation emissions, and they are likely to continue. Therefore, the EU issued a directive to include aviation in the EU ETS, which was published January 13, 2009. The intention is for the EU ETS to serve as a model for other countries considering similar national or regional schemes and to link these to the EU scheme over time. This way, the EU ETS can form the basis of a wider global action.

There is also a new proposal to reduce the CO_2 emissions from passenger cars. On December 19, 2007, the European Commission adopted legislation to reduce the average CO_2 emissions of new passenger cars, which account for about 12 percent of the European Union's carbon emissions. The proposed legislation is to improve the fuel economy of cars and ensure that average emissions from the new cars do not exceed 120 g/km of CO_2 through an integrated approach.

The Commission's proposal will reduce the average emissions of CO_2 in the EU from 160 g/km to 130 g/km in 2012—a 19 percent reduction of CO_2 emissions. This will make the EU a world leader in the production of fuel-efficient cars. Customers will benefit from fuel savings. From 2012, manufacturers will have to ensure that the cars they produce are meeting emissions standards. In addition, the curve is set so that heavier cars will have to improve more than lighter cars. Manufacturers' progress will be measured each year.

The EU also warns of the effects of climate change and the various adaptations that must take place to prepare for them. The EU stresses the importance of putting adaptation plans in place to soften impacts on society and the economy, including on water, agriculture, forestry, industry, biodiversity, and urban life. They also acknowledge that the impacts of climate change will hit locally and regionally in different ways and that adaptation measures will have to be planned out at local, regional, and national levels. To solve these issues and answer appropriate questions, there is currently an ECCP working group putting together an impact and adaptation plan, dealing with water resources, marine resources, coastal zones, tourism, human health, agriculture, forestry, biodiversity, energy infrastructure, and urban planning issues.

The *International Herald Tribune* reported on March 9, 2007, that the EU drafted an agreement that would make Europe a world leader in fighting global warming, but also compromised by allowing some of Europe's most polluting countries to limit their environmental goals. The draft agreement committed the EU to reduce GHG emissions by 20 percent by 2020 and required the EU to obtain one-fifth of its energy from renewable energy resources such as wind and solar energy, as well as fuel 10 percent of its cars and trucks with biofuels made from plants. Under pressure from several of the former Soviet bloc countries, however, which currently rely heavily on cheap coal and oil for their energy and fought changing to more costly environmentally friendly alternatives, the EU agreed that individual targets would be allowed for each of the 27 EU members to meet the renewable energy goal. Unfortunately, that means eastern Europe's worst polluters in the fastest-growing economies will most likely face the least stringent targets compared to their western counterparts. Many of the eight former communist nations that joined the EU in May 2004 are significantly behind the rest of the Union in developing renewable energy. Poland, for example, currently derives more than 90 percent of its energy for heating from coal.

In response to the agreement in general, however, the European Commission president, José Manuel Barroso, called the measures "the most ambitious package ever agreed by any institution on energy security and climate change," and expressed hope that they would spur the world's biggest polluters, including the United States, China, and India, to take similar action.

International Carbon Action Partnership

The International Carbon Action Partnership (ICAP), formed in October 2007, is a coalition of European countries, U.S. states, Canadian provinces, Australia, New Zealand, Tokyo Metropolitan government, and Norway formed to fight global warming. The international and interregional agreement was signed in Lisbon, Portugal, on October 29, 2007, by U.S. and Canadian members of the Western Climate Initiative, northeastern U.S. members of the Regional Greenhouse Gas Initiative, members of the European Union and the European Commission, Australia, Tokyo Metropolitan government, Norway, and New Zealand.

ICAP is designed to open lines of communication for sharing valuable information, such as research, effective policy initiatives, lessons learned, and new developments. By working together to establish similar design principles, ICAP partners are ensuring that future market systems, in conjunction with regulation in the form of enforceable caps, will boost worldwide demand for low-carbon products and services, provide a larger market for innovators, and achieve global emissions reductions at the fastest rate and lowest cost possible. The partnership supports the current ongoing efforts undertaken under the United Nations Framework Convention on Climate Change (UNFCC). ICAP is working toward finding global solutions by:

1. monitoring, reporting, and verifying emissions and working to determine reliable sources appropriate for inclusion in a globally linked program

2. encouraging common approaches and pushing partners to expand the global carbon market
3. creating a clear price incentive to innovate, develop, and use clean technologies
4. encouraging private investors to choose low-carbon projects and technologies
5. providing flexible compliance mechanisms that ensure reliable reductions at the fastest pace and lowest cost

According to UK prime minister Gordon Brown, "The launch of the International Carbon Market Partnership is a truly significant step forward in the global effort to combat climate change. Building a global carbon market is fundamental to reducing greenhouse gas emissions while allowing economies to grow and prosper. Trading emissions between nations allows us all to reach our greenhouse gas targets more cost effectively. And it therefore allows us to reduce emissions more than we could by acting alone."

Governor Jon Corzine of New Jersey commented, "My background as the former head of Goldman Sachs has given me a unique perspective on many market-based solutions to important public problems, such as environmental degradation. But it is my life in public service that has helped me understand that it will take the courage and commitment of a core set of leaders, like those of us gathered today, to drive implementation of smart, feasible, and measurable policies needed to address an issue as urgent as global warming."

Former governor Eliot Spitzer of New York said, "Global warming is the most significant environmental problem of our generation, and by establishing an international partnership, we are taking the vital steps to address this growing concern. In the absence of federal leadership, New York is implementing a greenhouse gas emissions trading program that will achieve a 16 percent reduction in power plant emissions by 2019. Today, we continue that work by joining the ICAP where we can begin working with our global partners, share experiences, and address issues of program design and compatibility, thereby strengthening our markets."

THE PROGRESS OF INDIVIDUAL COUNTRIES

Several of the world's countries have already made significant progress toward reducing their GHG emissions. In order to keep the Earth in a reasonable facsimile of what we know today, it will take the concerted effort of every nation on Earth. The scanty progress accomplished so far is discussed below.

Iceland

For the past 50 years, Iceland has been decreasing its dependence on fossil fuels by tapping the natural power found within its natural resources. Its waterfalls, volcanoes, geysers, and hot springs have long provided its inhabitants with abundant electricity and hot water. Today, virtually 100 percent of the country's electricity and heating comes from domestic renewable energy sources—hydroelectric power and geothermal springs. The country is still dependent, however, on imported oil to operate their vehicles and fishing fleets. It is so expensive to import that the cost is roughly eight dollars a gallon (two dollars a liter) for gasoline.

As of September 2007, Iceland ranks 53rd in the world in GHG emissions per capita, according to the U.S. Department of Energy's Carbon Dioxide Information Analysis Center. Professor Bragi Árnason of the University of Iceland has suggested using hydrogen to power the nation's transportation. Hydrogen is a product of water and electricity, and as he points out, "Iceland has lots of both." He further comments, "Iceland is the ideal country to create the world's first hydrogen economy."

His suggestion caught the attention of car manufacturers who are now using Iceland as a test market for their hydrogen fuel cell prototypes. One car that is receiving attention is the Mercedes Benz A-class F-cell—an electric car powered by a Daimler AG fuel cell.

Ásdis Kristinsdóttir, project manager for Reykjavik Energy says, "It's just like a normal car, except the only pollution coming out of the exhaust pipe is water vapor. It can go about 100 miles (161 km) on a full tank. When it runs out of fuel the electric battery kicks in, giving the driver another 18 miles (29 km)—hopefully enough time to get to a refueling station. Filling the tank is similar to today's cars—attach a

Sitting strategically on tectonic plate boundaries, Iceland has an abundance of geothermal energy that it can tap as a major energy source. *(Ásgeir Eggertsson)*

hose to the car's fueling port, hit 'start' on the pump, and stand back. The process takes about five to six minutes."

In 2003, Reykjavik opened a hydrogen fueling station to test three hydrogen fuel cell buses. The station was integrated into an existing gasoline/diesel fueling station. The hydrogen gas is produced by electrolysis—sending a current through water to split it into hydrogen and oxygen. The public buses could run all day before needing refueling. They calculated that Reykjavik would need five additional refueling stations; the entire nation will need just 15 refueling stations.

They expect that by the end of 2007, 30 to 40 hydrogen fuel cell cars will be driving on Reykjavik roads. Fuel cell cars are expected to go on sale to the public by 2010. The involved carmakers have promised they will keep costs down and the Icelandic government will offer its citizens tax breaks for driving them. Árnason figures it will take an additional 4 percent of power to produce the hydrogen. Once Iceland's vehicles are converted over to hydrogen, the fishing fleets will follow. He predicts Iceland will be completely fossil fuel–free by 2050. He said, "We are a very small country but we have all the same infrastructure of big nations. We will be the prototype for the rest of the world."

Iceland is also actively involved in carbon sequestration research. Icelandic, U.S., and French scientists have been studying chemical weathering and water/rock interactions for decades. They are interested in using Iceland as a location for carbon sequestration because the country's geologic formations are ideal for it and Icelanders' extensive knowledge of geothermal energy makes them good candidates for understanding chemical reactions between gases at the Earth's depths.

Sigurdur Reynir Gislason, a research professor of geology at the University of Iceland, said, "We hope to show the world in this pilot study that a natural process can be used to transform CO_2 emissions into a solid state and to safely store them underground for thousands, if not millions, of years. We also believe this process could not only be possible in Iceland, but in other countries that also have basaltic rocks."

Eileen Claussen, president of the PEW Center on Global Climate Change in Arlington, Virginia, said she is encouraged by such projects. "The PEW Center, along with many others, believe that carbon capture and storage underground in geological formations can be a significant part of the solution to climate change. Investment in these technologies illustrates the magnitude of the challenge and the lengths people are willing to go in order to change the dangerous path we're on."

Norway

Norway is another country involved in a pioneering effort to store CO_2 through carbon capture and storage. They have designated four separate sites: Sleipner, Snøhvit, Mongstad, and Kårstø. Since 1996, 1.1 million tons (1 million metric tons) of CO_2 from the Sleipner Vest oil field in the North Sea has been separated from the gas production and stored in Utsira (a geological formation) 3,280 feet (1,000 m) beneath the seafloor. Due to environmental concerns of leakage, the CO_2 storage facility is closely monitored. Several nations, supported by the European Union, have been involved in direct research and monitoring of this storage project, and they have developed prediction methods for the movement of the CO_2 spanning many years into the future. The resulting data is able to pinpoint the exact subsurface location of the CO_2 plume and confirm that the CO_2 is indeed confined securely within the designated storage reservoir.

The Snøhvit project began actively storing CO_2 on April 24, 2008, in an underground storage system. Natural gas, NGL, and condensate flows from the gas field in the Barents Sea. Up to 772,000 tons (700,000 metric tons) of CO_2 are separated annually from the natural gas and reinjected and stored in a formation 8,530 feet (2,600 m) under the seabed.

Mongstad, Norway, has plans to host the largest crude oil terminal and refinery. The Norwegian government and the oil company StatoilHydro has signed an agreement to establish a full-time CO_2 carbon capture and storage operation to offset a new gas-fired plant at Mongstad (Norway's largest crude oil terminal and refinery). The project will be completed in two phases: The first phase will cover construction and operation of the Mongstad CO_2 capture testing facility, which will be operational in 2011. The test facility will be able to capture at least 110,000 tons (100,000 metric tons) per year. Phase two will be full-scale capture of approximately 1.4 million tons (1.3 million metric tons) of CO_2 per year. This project is expected to be finished by the end of 2014.

In Kårstø—an existing area where carbon storage technology is already in existence—storage capacity will increase tenfold through a retrofit in 2011/2012. It will then capture and store approximately 1.1 million tons (1 million metric tons) of CO_2 each year.

Japan

According to a *USA Today* article of June 6, 2006, Japan hopes to cut back their GHG emissions and fight global warming with a plan to pump CO_2 into underground storage reservoirs rather than release it into the atmosphere. Fighting global warming is a top priority for Japan.

They release 1.3 billion tons (1.2 billion metric tons) of CO_2 each year into the atmosphere, making them one of the world's top polluters.

According to Masahiro Nishio, an official at the Ministry of Economy, Trade and Industry, Japan is planning to bury 200 million tons (181 million metric tons) of CO_2 a year by 2020, which will cut their emissions by one-sixth.

Carbon capture and storage (CCS) is a process whereby CO_2 is captured from factory emissions and pressurized into liquid form, then injected into underground aquifers, existing gas fields, or existing natural gaps between rock strata. The process is still under scientific investigation, although there is an experimental one being conducted in joint partnership between the U.S. Department of Energy (DOE), the Canadian government, and private industry. It began in 2005 and involved piping CO_2 from the Great Plains Synfuels plant in Beulah, North Dakota (a by-product of coal gasification), to the Weyburn oil field in Saskatchewan, Canada. In comparison, the proposed project in Japan is much larger.

According to Nishio, "Underground storage could begin as early as 2010, but there may still be hurdles to overcome. Capturing carbon dioxide and injecting it underground is prohibitively expensive, costing up to $52 a ton. Under the new initiative, the ministry aims to halve that cost by 2020. We have much to study in development."

Safety concerns also need to be addressed to ensure that earthquakes or rock fissures do not allow a sudden release of millions of tons of CO_2 into the atmosphere. The IPCC estimates that if CO_2 is stored properly and safely, it should remain stable for up to 1,000 years. Japan will begin their program by capturing CO_2 from their natural gas fields. Then, as they get the technology and program running systematically, they will also include CO_2 from steel mills, power plants, and chemical factories.

Nations Working toward Sustainability

An organization called the International Council for Local Environmental Initiatives (ICLEI) was established in 1990 as an international association committed to helping governments achieve sustainable development and mitigate and adapt to global warming. The ICLEI provides technical consulting, training, and information services tailored to countries' needs. They have worked with countries worldwide, such as in the examples discussed below.

Through ICLEI, farmers in the agricultural areas around Blantyre, Malawi, are currently changing their agricultural practices to support crops that need less water and nurture the soil. At the national level, the government has begun to increase the nation's grain reserve, anticipating more droughts and flooding in the years to come. It is also constructing a new dam in view of predicted future drought. The government is taking a proactive role in identifying measures it will need to take within the next three years in order to prepare itself for, and adapt to, climate change.

In Sapporo, Japan, ICLEI is involved in a project called Warm-Biz. This is a national program geared toward energy conservation. Run by the Japanese Ministry of the Environment, its purpose is to encourage people to wear more clothing to work to compensate for temperature settings being reduced by two degrees. In a pilot test program in Sapporo, 96.7 percent of the respondents supported the program overall and the citizens there learned that energy-efficiency programs offer one of the best ways to reduce global warming pollutants.

In Australia's Shire of Yarra Ranges, they have pledged to become carbon neutral. They have identified a range of innovative measures that significantly reduce their CO_2 consumption. Some of the positive measures they have enacted include adopting a climate change commitment that includes exceeding the Kyoto Protocol targets, becoming carbon neutral by reducing GHG emissions to a level 30 percent below 2000 levels by the year 2010, purchasing renewable energy certificates to offset emissions from street lighting, and offsetting council fleet emissions by planting 60,000 trees and progressively reducing energy consumption.

Residents of Canada Bay, Australia, are building a water mining plant that will save drinking water. The plant will save up to 44 million gallons (165 million l) of drinking water each year by providing recycled water for the city's fields, golf courses, and parks. The plant will work by purifying wastewater, using mechanical methods and minimal chemicals to produce high-quality treated water.

London is planning to cut GHG emissions by 60 percent with the next 20 years. Their plan aims to reduce emissions at the local government, industrial, and business levels. Individual elements of their plan include awarding green badges of merit for local businesses adopting reduction strategies, offering subsidies to homeowners to insulate their homes, and switching one-fourth of the city's power supply from the old and inefficient national grid to locally generated electricity using combined heat and power plants.

According to former London mayor Ken Livingstone, speaking of London's Climate Change Action Plan, "Londoners don't have to reduce their quality of life to tackle climate change, but we do need to change the way we live."

On November 17, 2007, in Valencia, Spain, UN Secretary-General Ban Ki-moon described climate change as "the defining challenge of our age." He also challenged the world's two largest GHG emitters—China and the United States—to "play a more constructive role." His challenge was delivered two weeks before the world's energy ministers met in Bali, Indonesia, to begin talks on creating a global climate treaty to replace the Kyoto Protocol when it expires in 2012.

The IPCC, which was awarded the Nobel Peace Prize (jointly with Al Gore) in October 2007, said the world would have to reverse the growth of GHG emissions by 2015 to prevent serious climate disruptions. According to Dr. Rajendra Pachauri, chair of the IPCC, "If there's no action before 2012, that's too late. What we do in the next two to three years will determine our future. This is the defining moment."

He also said that since the IPCC began its work five years ago, scientists had recorded "much stronger trends in climate change," like a recent melting of Arctic ice that had not been predicted. "That means you better start with intervention much earlier."

One of the major differences with the IPCC's fourth assessment report (released in 2007) over previous ones was that the data had not been softened, diluted, and sifted through. It was direct and to the point. It was the first report to acknowledge that the melting of the Greenland ice sheet from rising temperatures could result in a substantive sea-level rise over centuries rather than millennia. It added a sense of critical urgency and importance never seen before in a report.

"It's extremely clear and is very explicit that the cost of inaction will be huge compared to the cost of action," said Jeffrey D. Sachs, director of Columbia University's Earth Institute. "We can't afford to wait for some perfect accord to replace Kyoto, for some grand agreement. We can't afford to spend years bickering about it. We need to start acting now."

"Stabilization of emissions can be achieved by deployment of a portfolio of technologies that exist or are already under development," said Achim Steiner, head of the UNEP. But he noted that developed countries would have to help poorer ones adapt to climate shifts and adopt cleaner energy choices, which are often expensive. Mr. Steiner emphasized that the report sent a message to individuals as well as world leaders: "What we need is a new ethic in which every person changes lifestyle, attitude, and behavior."

Global Warming, Human Psychology, and the Media

The media has an enormous influence on what the public hears about. It is the media that disseminates information, through newscasts, magazines, newspapers, the Internet, giving them an unparalleled opportunity not only to inform the public of the latest issues, but also to play a role in how that information is perceived. Another component that contributes to how information is received is different for each person on Earth and is based on preferences, perceptions, and beliefs that are influenced by psychology and value systems. This chapter delves into these issues for a look at the sometimes-subtle forces at work shaping people's opinions about highly controversial subjects, such as global warming.

HUMAN PSYCHOLOGY AND CULTURAL VALUES

According to Dr. H. Steven Moffic, a professor of psychiatry and behavioral medicine at the Medical College of Wisconsin, "Global warming is a concept that everyone hears about, but many are slow to respond to. The problems and risks of global warming seem to be far in the future—they might be 25 or 50 years away—so why would people pay attention to those issues when there are so many day-to-day problems to deal with?"

Dr. Moffic believes the ability to ignore global warming is very human. "Our brains in many ways have not evolved much from when humans started to develop thousands of years ago. We are hardwired to respond to immediate danger—we call this the 'fight or flight response'—but there is no similar mechanism that alerts us to long-term dangers."

He believes that these reactions are just part of human nature. "People are so preoccupied with immediate problems like jobs and health and the economy that it's hard to pay attention to global warming, and to willingly take on another challenge.

"The issue of how much humans contribute to the cause of global warming may also contribute to why we tend to ignore its impact. Who wants to believe they might be guilty for contributing to a problem that could destroy the Earth?"

In order to put the issue in perspective, Dr. Moffic suggests everyone identify and do simple things that do not require big changes. He believes that each individual can have a large effect on others and, through example, influence others to take action. He also suggests that everyone "try to make global warming a more immediate issue—whether it is thinking about your kids, grandkids, the future of the whole Earth, or your health. Try to think about ways in which this issue is important to you right now."

In work done by Elke U. Weber at the Center for Research on Environmental Decisions at Columbia University on why the subject of global warming has not scared more people yet, she attributes it to universal characteristics of human nature. According to Weber, behavioral decision research over the last 30 years has given psychologists a good understanding about the way humans respond to risk; specifically in the decisions they make to take action to reduce or manage those risks. One of the biggest motivators to respond to risk is worry. When people are not alarmed about a risk or hazard, their tendency is not to take precautions.

Weber points out that with the issue of global warming, personal experiences with notable and serious consequences are still rare in many regions of the world. In addition, when people base their decisions on statistical descriptions about a hazard provided by others, it is not a big enough motivator for action.

An example of this can be seen in a scenario such as the rapid rise in the price of gasoline in 2008. When prices skyrocketed at the pumps, it caught the public's attention and raised an immediate interest in hybrid cars, alternative fuels, and using public transportation, because the consumer was hit hard financially. Then, when gasoline prices dropped again, consumers thought less about energy conservation and alternative fuels because they were no longer immediately suffering the direct consequences. Human nature dictates that if something negative happens elsewhere in the world, the mindset of an individual is "it only happens to others."

The stark reality about global warming is the inertia it engenders. Other locations may be suffering through droughts (such as Africa) or sea-level rise (such as the Pacific or Caribbean Islands), but people think it won't happen in the United States. Sadly, when it eventually does, it will already be too late. And just like it is human nature to procrastinate when an immediate threat is not looming, eventually the public will be caught in the mindset: "I wish I had done something about it sooner."

Weber also believes that the reason people tend to avoid taking action against long-term risks is related to two psychological factors: the finite pool of worry hypothesis and the single action bias. The finite pool of worry hypothesis posits that people can only worry about so many issues at one time, and of the issues they worry about they are prioritized from greatest to least. Generally, the greatest worries are those most directly affecting their lives at the moment. As an example, Weber pointed out that the finite pool of worry was demonstrated by the fact that in the United States there was a rapid increase in concern about terrorism after the attacks on 9/11. Because of the intense focus on terrorism, other important issues—such as environmental degradation or restrictions on civil liberties—took an immediate backseat.

The single action bias is described by Weber as follows: "Decision makers are very likely to take one action to reduce a risk that they encounter and worry about, but are much less likely to take additional steps that would provide incremental protection or risk reduction. The single action taken is not necessarily the most effective one, nor is it the same for different decision makers. However, regardless of which single action is taken first, decision makers have a tendency to not take any further action, presumably because the first action suffices in reducing the feeling of worry or vulnerability."

Weber concludes that based on behavioral research over the past 30 years attention-grabbing and emotionally engaging information interventions may be required to ignite the public concern for action in response to global warming.

A country's cultural values also play a significant role in public perception—and reaction—to global warming. An individual's values promote public action on issues such as civil rights, feminism, the jobs and social justice movements, the peace movement, the organic food and alternative health care movements, and the environmental movement. According to the State of the World Forum, these movements have gained strength over the past 50 years. In the United States alone, they estimate that more than 50 million people support some sort of groups based on personal values, such as those seeking to protect the environment. The numbers continue to grow and in Europe are even more numerous.

These movements have power over political decision makers. Organizations with influence include Defenders of Wildlife, World Wildlife Fund, and Union of Concerned Scientists. For a listing of such organizations, see the Appendix.

THE POWER OF THE MEDIA

Reporting about global warming by the media has run the gambit in recent years. Since there are many points of view, the question is where does the truth lie. Reports and stories concerning global warming have ridiculed scientists and environmental groups. Reports have shown big businesses and countries (such as the United States) openly challenging the facts of climate change. Industries, such as oil companies, have accused the media of misinforming the public about the ill effects of burning fossil fuels. Other news stories have accused the Bush administration of silencing critics, including leading government climate scientists who have warned the public openly of the consequences of global warming.

As further reports about global warming continue to reveal a bleaker future, some are concerned that it will encourage fear tactics from environmentalists, whitewashing by some business interests, and a show by governments to illustrate reductions in emissions.

A few media reports claim global warming to be a fraud; still others claim it is simply a cause designed to harm the U.S. economy and make the United Nations more powerful. Others say it is driven by academia and the simple desire of climate scientists to make a lot of money by using fear as a tool to earn more research grants.

All of this misinformation presents a challenge. A trend that has emerged is that the mainstream media in recent years has turned toward reporting actions and solutions. But there does seem to be a fine line on what the public expects. Some global warming researchers have expressed concern that too much reporting will lead to climate fatigue whereby the public will become desensitized to the issue. Others feel that the media should be used as an educational tool, that there is so much potential to educate the general public in ways that are not fatiguing. As an example, consider the two photos. The billboard advertises a pickup truck dependent on fossil fuels and not rated with high fuel efficiency, but the advertising has emotional appeal by suggesting the luxury, comfort, and status that will be bestowed upon the buyer of the vehicle. When the public looks at this type of advertisement, they are not reminded of global warming issues or the health of the environment and future generations.

The movie poster sends an entirely different message. Focused on the Earth and those who live on it, it communicates very well the connections between life on land, in the oceans, and the overall connection to everything on Earth. This type of media representation serves as not only entertainment but as a strong positive approach to public education applicable to people of all ages. Instead of causing environmental fatigue, it sparks environmental interest through its creative storyline and breathtaking photography, giving the viewer a glimpse of the diversity and fragility of life on Earth that they probably would never see otherwise.

Commercial media is geared to appeal to a consumer's ego, desires, comfort level, and status. By presenting a product in this way, it is much easier to generate personal interest, because making a sale is the goal. If a global warming scientist were to recreate this advertisement it would read much differently and carry a much different message. *(Nature's Images)*

KEEPING A JOURNALISTIC BALANCE

Journalistic balance—giving both sides of an issue—is an important concept. The organization Fairness and Accuracy in Reporting (FAIR) states that a new study found that in U.S. media coverage of global warming, superficial balance—telling both sides—can actually be a form of informational bias. An example of this concerns global warming. As the IPCC, for example, has reiterated that human activities have had a discernible influence on the global climate and that global warming is a serious problem that must be addressed immediately, the media, in the name of balance, have given disproportionate air play to the small group of global warming skeptics and allowed them to have their views greatly amplified.

When reading reports from the media, it is important to clearly note who is being interviewed and whether or not the source is reputable and noteworthy. For example, the IPCC is reputable. It consists

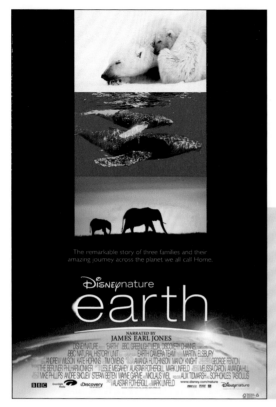

The media has the power to contribute to public education about the environment in a positive manner. One highly successful way is through entertainment. With good narrative and photography, a strong impression can be made to individuals, strengthening public involvement around issues such as doing their part in solving the global warming problem. *(Disney)*

of top scientists from around the globe, and they employ a decision-by-consensus approach. To back their reputabilty, D. James Baker, administrator of the U.S. National Oceanic and Atmospheric Administration (NOAA) and undersecretary for oceans and atmosphere at the Department of Commerce under the Clinton administration, has said, "There's no better scientific consensus on this or any issue I know—except maybe Newton's second law of dynamics."

In 1996, the Society of Professional Journalists removed the term *objectivity* from its ethics code. Today, the trend seems to lean more toward fairness, balance, and accuracy. Journalists today are taught to identify the most dominant, widespread position and then tell both sides of the story. Robert Entman, a media scholar, says, "Balance aims for neutrality. It requires that reporters present the views of legitimate spokespersons of the conflicting sides in any significant dispute and provide both sides with roughly equal attention.

"Balanced coverage does not, however, always mean accurate coverage. In terms of global warming, 'balance' may allow skeptics—many of them funded by carbon-based industry—to be frequently consulted and quoted in news reports on climate change. When the issue is of a political or social nature, fairness—presenting arguments on both sides with equal weight—is a fundamental check on biased reporting. But this tactic can cause problems when it is applied to scientific issues. It encourages journalists to present competing points of view on a scientific question as though they had equal scientific weight, when they, in fact, do not. This is what has happened with global warming. Some media has let the skeptics have too much voice and has enabled them to confuse the public and distort the seriousness of the problem.

"By giving equal time to opposing views, the major mainstream media has significantly downplayed scientific understanding of the role humans play in global warming. While there is a value to presenting multiple viewpoints, it does the reader a disservice when scientific findings that have been agreed upon by many of the world's top scientists are presented next to (and with equal weight) the opinions of a few skeptical scientists. Confusing to the reader, they no longer know what to believe. Situations like this slow down the constructive progression of global warming research."

An example of this can be seen in a *New York Times* article from April 10, 2009. Marc Morano sponsors a Web site (ClimateDepot.com) dedicated solely to the downplay of global warming. His chief goal is to debunk global warming as a serious issue. Kert Davies, the research director for Greenpeace, commented that he would "like to dismiss Mr. Morano as irrelevant, but could not. He is relentless in pushing out misinformation. In denying the urgency of the problem, he definitely slows things down on the regulatory front. Eventually, he will be held accountable, but it may be too late."

As scientists who are actively involved in global warming research look into Mr. Morano's claims, they say "he may be best known for compiling a report listing hundreds of scientists whose work he says undermines the consensus on global warming. Environmental advocates, however, say that many of the experts listed as scientists on Morano's

Web site have no scientific credentials and that their work persuaded no one not already ideologically committed."

One of Morano's recent reports entitled "More than 700 International Scientists Dissent over Man-Made Global Warming Claims" was far from balanced." Kevin Grandia, who manages Desmogblog.com, which describes itself as dedicated to combating misinformation on climate change, says the report is filled with so-called experts who are really weather broadcasters and others without advanced degrees. Mr. Grandia also said Mr. Morano's report misrepresented the work of legitimate scientists. Mr. Grandia pointed to Steve Rayner, a professor at Oxford, who was mentioned for articles criticizing the Kyoto Protocol. Dr. Rayner, however, in no way disputed the existence of global warming or that human activity contributes to it, as Morano's report implied. In e-mail messages, he had asked to be removed from the Morano report, but his name was not, it was published with it included. When asked about it, Morano replied that he had no record of Dr. Rayner's asking to be removed from the list and that the doctor must "not be remembering this clearly."

In cases like these, it is imperative that any information obtained about global warming—or any scientific issue, for that matter—be looked at critically and its validity assessed as to its scientific soundness and quality.

SCIENTISTS' MINDSETS AND DATA CHANGE

One way the media has negatively impacted the advancement of global warming research is to attack scientists when they have changed their theories or their positions on a scientific viewpoint. For example, the media brought up a theory postulated back in the 1970s that did not pan out and allowed outspoken critics to use it in an attempt to diminish the reputations of scientists today. Several mainstream media sources republished the stories from the 1970s about a coming age of global cooling and the climate disaster it would trigger. Because this nearly 40-year-old theory never panned out, some skeptics have said global warming will not pan out either. But scientists say that is an unfair comparison.

Dr. William Connolley, a climate modeler for the British Antarctic Survey, says that "Although the theory got hype from the news media in the '70s, it never got much traction within the scientific community; but that new data and research over the decades have convinced the vast majority of scientists that global warming is real and under way."

The issue in the 1970s centered around the possibility that nearly three decades of cooling experienced in the Northern Hemisphere since World War II might be the beginning of a new ice age. Data suggested that perhaps the huge increase in dust and aerosols from pollution and development might be stepping up the cooling process. The investigation did not last long, however, because temperatures began to rise again and the issue was abandoned. Today, improved climate methodologies have revealed that although aerosols did have a cooling effect, CO_2 and other GHGs were more potent in bringing about atmospheric change on a global scale. Improvements in technology over recent years have greatly aided the advancement and accuracy of scientific research, which continues to evolve and improve.

Back to the issue of climatologists changing their minds, however. R. Stephen Schneider, a professor in the department of biological sciences at Stanford University and a senior fellow at the Center for Environment Science and Policy of the Institute for International Studies, says, "Scientists are criticized by global-warming skeptics for making new claims and revising theories, as if we are required to stay politically consistent. But that goes against science. We must allow for new evidence to influence us.

"For some, the original speculation was that dust and aerosols would increase at a rate far beyond CO_2 and lead to global cooling. We didn't know yet that such effects were so regionally located. By the mid-1970s, it was realized that greenhouse gases were perhaps more likely to be shifting climate on a global scale."

Dr. Connolley stated, "Climate science was far less advanced in the 1970s, only beginning in a way, and ideas were explored in a tentative way that has later been abandoned."

This represents an inherent issue of science in general. As additional knowledge is gained about a subject, processes and outcomes of phenomena may change. Scientists need to remain open-minded and objective. If they do not remain open-minded, they will miss critical pieces of scientific information and possibly risk the outcome of a scientific breakthrough.

One thing remains clear, however. The media, if used correctly, have an enormous potential to guide the public and can play a significant role in helping people understand the science, the relevant issues, and the options for a better future.

The Stand on the Debate

Global warming is one of the most controversial issues today. There are not only extreme right and left points of view, but there are gradations of every degree in between. The issue has caused heated debates among the world's most respected climate scientists. It generated controversy back when Jean-Baptiste Fourier began making connections with the Earth's natural *greenhouse effect* and heat properties, and controversy and tension still surround the subject today, even though there have been many scientific breakthroughs that have provided compelling evidence of its existence.

The controversy spans many platforms—scientific, political, economic, environmental, cultural, and ideological—and affects every member of society regardless of where they live on Earth. It also involves a blend of changes that are (1) natural and (2) anthropogenic (human-caused) working on multiple time intervals, some short-term nested within long-term changes, some part of a predictive cycle, others on their own time cycle, and still others unpredictable. What may seem clear and logical to some may seem like chaos to others trying to make sense of the Earth's climate—certainly one of the most complex systems in existence. And a final component that makes this issue so difficult is its personal scale—it is not a problem a single invention, government, or wealthy research institution can fix—it will take every human on the face of this planet making permanent sacrifices and commitments for the good of all. It is not a spectator issue that will merely require one to turn the TV on to check on its progression, it will take participation and personal commitment—there is no place to hide and no exceptions. This chapter illustrates the present-day opinions and stances taken on this issue.

UNDERSTANDING MODERN CLIMATE

Throughout the 1970s, multiple opinions existed about the climate, and no strong consensus rose above the confusing jumble of theories as to whether the Earth's climate was really warming or cooling. There was a multitude of data collected, but not all of it was reliable. The Goddard Institute for Space Studies (GISS)—a group funded by NASA—began sifting through the enormous amounts of data, discarding that which was not reliable and using that which was. Dr. James E. Hansen, one of the most notable experts on global warming today, led this group. They were able to analyze the data computer modeling programs they had developed for data pertaining to both the Northern and Southern Hemispheres.

According to Dr. Hansen, in 1981, "The common misconception that the world is cooling is based on Northern Hemisphere experience to 1970." He pointed out that around the same time that meteorologists had noticed the cooling trend in the weather records, they began to reverse direction once again. According to Hansen, from the low point in the mid-1960s to 1980, the Earth's atmosphere had actually warmed 0.3°F (0.2°C). He was able to determine that the cooling trend in the '60s and '70s was due to volcanic eruptions, changes in the Sun's energy output, and an increase in pollution in the industrialized portions of the Northern Hemisphere.

Unfortunately, the attention the temporary cooling trend received from scientists, the media, and the general public served to throw doubt and skepticism toward the theory of the enhanced greenhouse effect and global warming. GISS's viewpoint, which they stated at the time, however, was that greenhouse warming had been masked during the '60s and '70s by "chance fluctuations in solar activity, volcanic aerosols, and increased haze from pollution." They also predicted that "considering how rapidly CO_2 was accumulating, by the end of the 20th century, carbon dioxide warming should emerge from the noise level of natural climatic variability."

The Climatic Research Unit at the University of East Anglia, operated by the British government, also analyzed the mass of climatic data and like NASA's GISS came to similar conclusions: A warming trend due to greenhouse gases would present itself clearly in the records by 2000.

Even with the endorsement of the world's two leading climate research institutions, many of the world's climate experts did not support the notion that the Earth's atmospheric temperature would continue to steadily warm from the 1970s forward. Doubt stemmed from the fact that reliable data only existed for the past 100 years, and within that time period had already shown a significant degree of variation. Many believed that future activity in either direction would merely be a "wobble" in the temperature. As shown in the illustration, however, from 1970 forward, it is clear that NASA/GISS and the British Climatic Research Unit were correct in their predictions.

By 1990, the National Oceanic and Atmospheric Administration (NOAA) National Climatic Data Center entered the picture. They held possession of the world's largest collection of historical weather records and were busy organizing all the data collected from the Weather Bureau and military services from the 1940s on. Thomas Karl led the team of scientists at NOAA, who carefully reviewed the statistics for world weather and climate.

As it turned out, the decade of the 1980s included four of the Earth's warmest years on record. Then, in the early 1990s, temperatures dipped downward again. NOAA, NASA, and the majority of climate scientists attributed the drop to the eruption of Mount Pinatubo in the Philippines. A major volcanic eruption, the ejection of particulates was so enormous it temporarily reduced atmospheric temperatures by blocking incoming solar radiation worldwide.

Once adequate precipitation had washed the volcanic particulates from the atmosphere, temperatures began rising once again, making 1995 the warmest year on record. Temperatures did not stop rising; 1997 was hotter than 1995, and 1998 quickly replaced 1997 as the hottest year ever on record, then after that, 2002 and 2003, and the trend continued. According to NASA/GISS, 2005 was the warmest year in over a century at that time. And it did not stop there.

NASA has determined that 2007 was the second warmest year globally—and the hottest year on record in the Northern Hemisphere. According to the Earth Policy Institute, "It is clear that temperatures around the world are continuing their upward climb. The global average in 2007 was 58.5°F (14.7°C), which makes it the second warmest year on record, only 0.05°F (0.03°C) behind the 2005 maximum. January 2007 was the hottest January ever measured, a full 0.38°F (0.23°C) warmer than the previous record. August was also a record for that month and September was the second warmest September recorded."

Extremely notable is the Northern Hemisphere for 2007. Temperatures averaged 59.1°F (15.0°C), by far the hottest year in the Northern Hemisphere since temperature records began being collected in 1880. This is also more than a degree warmer than it was during the 1951–1980 time interval, showing recent marked warming. As scientists have compared this data to the ancient paleo records (such as tree rings and ice cores), this is also warmer than it has been at any time in the past 1,200 years.

One of the most interesting things about 2007 being such a warm year was that there were several natural conditions present during that year that should have cooled the climate. That year experienced a moderate La Niña, which should have countered warming effects. The solar intensity was also slightly lower than average because the 11-year solar sunspot cycle was at a minimum. According to the Earth Policy Institute, "The combination of these factors would normally produce cooler temperatures, yet 2007 was still one of the warmest years in human history." They believe the high temperatures are attributed to the warming effect of increased greenhouse gas concentrations causing global warming.

Another interesting component is that several areas worldwide experienced extreme weather. In southeastern Europe, for example, temperatures climbed as high as 113°F (45°C) in a heat wave that killed up to 500 people. Japan also experienced extreme heat waves, with temperatures reaching 106°F (41°C). Other areas, such as Greece and the American West (Utah, Colorado, Nevada, California, New Mexico, Idaho, and Wyoming) experienced extreme high temperatures and drought, which proved a deadly combination and contributed to massive wildfires during the summer and fall. Other areas experienced record-breaking amounts of rain. England and Wales suffered through widespread flooding, creating $6 billion in damage. South Asia also saw record-breaking flooding, which killed over 2,500 people. Floods in Africa caused hundreds of thousands of people to lose their homes and farmlands, leaving them with nothing.

The World Meteorological Organization stated that "There were indications that the 10 years from 1998–2007 were the hottest decade on record. The Met Office Hadley Centre said the top 11 warmest years have all occurred in the last 13."

Because climate change has regional variations so that different geographic locations may experience different degrees of temperature change, when climatologists looked at the climate system globally, by the late 1990s the majority of scientists generally acknowledged some degree of global warming. There was still a minority of very outspoken critics, however. They argued over global warming for several reasons. Some argued that the urban heat effect from cities was still skewing test results and wrongly making the climate look warmer than it actually was, even though scientists at both NOAA and NASA had thoroughly analyzed all past data and accounted for any additional heat being contributed due to industrialization and urbanization and removed its effect from the temperature calculations. Critics also refused to acknowledge the existence of proxy data collected far away from urban areas, such as tree rings, coral, and ice cores, which clearly showed long-term warming trends were underway.

Another major point critics focused on was temperature data acquired from satellites. In 1979, satellites were deployed to orbit Earth and collect continuous climate data. This represented a breakthrough as a reliable, continual source of global climate data. Critics, however, discounted its relevance because they claimed the instruments measured the temperatures of the middle heights of the atmosphere, not the Earth's surface, and at the middle heights there had been a slight decrease in temperature.

This was embarrassing to the climate scientists developing climate models, because their models had actually predicted that the midatmospheric levels would show a warming, but the creation of climate models was in its infancy and there was still much to learn about climate behavior and how to build incredibly complicated, sensitive models that needed to take thousands of variables into consideration and provide accurate outcomes.

Climate models have evolved over time and, interestingly, one of the things that scientists have cleared up is that once better analytical capabilities were developed, scientists were able to determine that the atmosphere's midlevel was warming just as the models had predicted they would.

As the warming trend continued, toward the end of the 1990s, enough indicators were present that the majority of scientists acknowledged that a universal warming was taking place. This decision was gained through ancillary data, such as winter snow cover melting earlier in the spring in the Northern Hemisphere, leaves budding earlier on trees in the spring, and a warming trend in the ocean's surface.

Therefore, with all this ongoing fluctuation and science's struggle to unravel all the complicated natural and man-made cause-and-effect relationships, it has made it difficult for the scientific community to come together and support a common viewpoint and come to a single agreement. Just as global warming will not affect every place on Earth the same way—some will experience drought, others flooding—the evidence is not universal either. Some exists as small changes, like flowers blooming two weeks earlier; others manifest as larger clues, like the spectacular collapse of huge ice shelves in Antarctica. The diversity of clues and the complexity and difficulty of predicting the climate have led some people to doubt the existence of global warming, while others are thoroughly convinced the problem needs urgent attention.

Unfortunately, it is the controversies between groups of opposing opinions that have partly caused such a delay in acting quickly in order to solve the problems associated with global warming. The next sections illustrate some of these controversies and heated issues.

According to the Union of Concerned Scientists (UCS), throughout the 1990s "Climate change has come to be accepted as one of the biggest, most complex scientific and political challenges the world has ever faced. Not possible to solve with simple solutions, it will remain a key problem of the next century and even longer."

According to UCS, one of the most promising developments is that scientific methods and data collection and analysis collection techniques have advanced recently. Climate science has matured, long-term observational data is available, analytical and computer technology has improved, and scientists are collaborating under the guidance of the IPCC. Even so, there are still distinct groups in the scientific community, political realm, media, and general public that are skeptical that global warming exists. These groups tend to be very outspoken in their protests and commonly seek the support of the general public as well as the U.S. Congress. There is also a newly emerging group of those called the middle group—those who believe global warming is both a natural process and human-induced in particular aspects. Their viewpoints lie somewhere between the extremes. And finally, there are the supporters of global warming that recognize the threat and are taking and supporting action to make a difference.

THE FAR RIGHT—SKEPTICS OF GLOBAL WARMING

The skeptics of global warming are those individuals and groups that do not believe that human-caused global warming is presently occurring, or that there is a danger of it occurring in the near future. Many of these skeptics have been very outspoken against the IPCC and its 2,500 or more scientists who have analyzed the worldwide climate data and determined the effects of global warming on nations' economies, cultures, traditions, and lifestyles.

According to SourceWatch, skeptics are somewhat predictable and usually will argue against the existence of global warming centered on the variations of four lines of thought:

1. Some skeptics claim there is a lack of conclusive evidence that global warming is actually happening right now.
2. Other skeptics say that any changes that are occurring in the weather right now are simply part of the Earth's natural cycles;

that the climate cycles naturally through warmer and colder periods regardless of what humans do to the environment.
3. Some say that even if humans are somewhat responsible for some of the climate changes that are occurring, the scale of the impact is not large enough to call for drastic, costly changes.
4. Skeptics also claim that it would be too expensive and difficult to make the suggested cuts in greenhouse gas emissions recommended by the IPCC.

The UCS has identified several methods that skeptics use to discredit the science behind global warming. The first strategy is to discredit the message about global warming. This is done by three basic methods:

Focus attention on scientific uncertainties rather than discoveries. With this method, skeptics exaggerate scientific uncertainties at the expense of solid, established scientific findings. Their goal is to convince the public and policy makers that no one needs to take any action now—it is okay to wait until climate change is a certainty. This encompasses attitudes such as "It can be taken care of sometime later, if at all," or "It is not my problem, anyway."

This happened with a report that discussed some uncertainties about comparing data collected at the Earth's surface versus that collected from a satellite. Skeptics used the mention of a discrepancy of the data as proof that global warming was not real. The actual study, however, went on to clarify that despite the temperature differences, there was still a substantial rise at the Earth's surface over the past 200 years. Simultaneously, the Greening Earth Society—a skeptic's organization funded by Western Fuels Association—released a statement about the same report but stated that "global temperatures have not been changing exactly as the models had predicted." They concluded that the report was proof that global climate model forecasts are unreliable indicators of future climate.

- Emphasizing and taking out of context selected findings to weaken the scientific conclusions. In this method, skeptics pick and choose from the scientific findings to support their case. Often they take findings out of context.

- Make false claims for the policy implications of scientific findings. Some skeptics undermine the calls for action based on convincing evidence by starting "Yes, but . . ." arguments to foster doubt. For example, in a proposed action to correct a pollution problem, instead of focusing on the benefits of cleaner air and better energy efficiency, they will focus on the economic burden. The goal with this strategy is to undermine and trivialize the proposed action. Skeptics also deliberately misconstrue scientists' findings and conclusions.

Discredit the messenger. In this second method, skeptics are not beneath name calling and attempt to turn global warming into a political issue by discrediting specific political figures. Often relying on pitting party against party, discrediting the messenger happened in a 1999 article in the *National Journal* with an attack on Al Gore. The political climate today is even more volatile due to the current global issues with OPEC, the rocketing prices of a barrel of oil, and pressure being put on the United States to take action against global warming and support the immediate implementation of renewable energy. Global warming was a major issue in the 2008 presidential election, and skeptics used opportunities like this to attempt to discredit global warming.

Discredit the process through which scientific results are achieved. An example of a third strategy occurred when an IPCC report was issued. Skeptics accused one of the lead authors of making unauthorized changes to a chapter after its acceptance by the IPCC. Skeptics claimed the report had been altered and that the chapter had been "cleansed" of all the discussions of scientific uncertainties.

Even though the IPCC responded to the allegations by saying the changes made were done to "improve its presentation, clarity, and consistency in accordance with the view both of scientists and delegates expressed at length during the meeting" and the IPCC verified that "the modifications did not change the bottom-line conclusion, nor were uncertainties suppressed," the skeptics did not let it go. They promoted the episode as unethical for months afterward in an attempt to lessen the integrity of the lead author and thereby invalidate all the contributing IPCC scientists' findings.

According to SourceWatch, another strategy is to magnify the counter-message. In this strategy, skeptics focus on positive aspects of global warming. For example, an article published in October 1998 in *Science* estimated how much carbon dioxide could be absorbed from the atmosphere by major terrestrial *carbon sinks*. In one study, carbon uptake in North America exceeded annual emissions. Skeptics took this piece of information out of context and focused on it, leading the public to believe the United States had no real role in combating global warming. According to Peter Huber in an April 1999 article in *Forbes* magazine: "If the estimate is right, we don't owe the rest of the world a dime on carbon emissions. They owe us. Americans recycle our carbon. If greenhouse gas is a problem at all, the rest of the world is the problem. America's the solution. Perhaps we could do even more. But the fact is, we're doing more than our share already."

A July 23, 1999, article in *Science* by R. A. Houghton later countered this message, clearly taken out of context. He dispelled their findings and illustrated that the net carbon flux related to U.S. lands offset 10 to 30 percent of the United States' fossil fuel emissions. Still in its infancy, research continues today on carbon flux issues.

The Greening Earth Society focused on carbon dioxide emissions being a "wonderful gift to the world's agricultural sector." They claimed that the world would be able to produce more food for growing populations, thereby eliminating hunger. What they left out of their analysis, however, was that additional CO_2 also leads to increased drought, water-stressed vegetation, vulnerability to insect pests, increased exposure to the spread of disease, and the additional risk of wildlife hazards.

Another strategy to push the counter-message is through the creation of skeptics' organizations. Once an organization is formed, there is nothing to prevent it from going to Capitol Hill to lobby against global warming. According to an article in *The Age* in June 2005, climate skeptics in Australia reported global warming to be merely a cyclical phenomenon that has occurred throughout the Earth's history, not a human-caused situation to be concerned about. Dr. Rob Carter at James Cook University believes the rising level of carbon dioxide has actually been good for agriculture, the proof of which has been in increased crop yields.

"Carbon dioxide is the best aerial fertilizer we know," he told the reporter. He also stated that "the Kyoto Protocol would cost billions, even trillions, of dollars and would have a devastating effect on the economics of countries that signed it. It will deliver no significant cooling—less than 0.3°F (0.2°C) by 2050. Climate has always changed and always will. The only sensible thing to do about climate change is to prepare for it."

One of the key issues skeptics have focused on is the discrepancies in temperature data taken from the atmosphere versus that taken from the ground. This has been a major arguing point for years. The atmospheric temperatures have not risen like ground temperatures have, leaving skeptics to promote the idea that these factors are not related. However, with a new study released by *Live*Science in May 2006, the temperature discrepancy has finally been resolved. In a report prepared by the U.S. Climate Change Science Program, the errors in the satellite and radiosonde data have been identified and corrected. Their findings also clearly indicate that human influences have been directly involved. They targeted the releases of gases such as carbon dioxide into the atmosphere from burning fossil fuels for transportation and industrial activities.

According to Thomas R. Karl, director of the National Climatic Data Center, "There are still some questions about the rate of the atmospheric warming in the Tropics, but overall the issue has been settled."

The final report concluded that:

- Since the 1950s, all data show the Earth's surface and the low and middle atmosphere have warmed, while the upper stratosphere has cooled. This trend also matches the computer models designed to portray the greenhouse effect.
- Radiosonde readings confirmed that the mid-*troposphere* warmed faster than the surface, which also agreed with the greenhouse model (a radiosonde is an instrument carried aloft by a balloon to transmit meteorological data by radio).
- Natural processes cannot account for the patterns of change documented during the last 50 years alone—it can only be explained with human interference added to the natural processes.

The following table depicts what some prominent skeptics are saying today.

\multicolumn{2}{c}{**What Skeptics of Global Warming Are Saying**}
SOURCE	COMMENT/ACTIONS
Richard S. Lindzen, professor at MIT	He is willing to bet the Earth's climate will be cooler in 20 years than it is today.
Sallie L. Baliunas, astrophysicist at Harvard-Smithsonian Center for Astrophysics	She believes that global warming is a hoax.
Exxon-funded skeptics	Since 1990, they have spent more than $19 million funding groups that promote global warming skepticism and $5.6 million to public policy organizations that publicly deny global warming and climate change.
Philip Stott, professor at University of London	He questions the knowledge of the IPCC.
Patrick J. Michaels, former professor at University of Virginia	He believes that global warming models are fatally flawed and, in any event, we should take no action now because new technologies will soon replace those that emit greenhouse gases.
Bjorn Lomborg, professor at University of Aarhus, Denmark	He wrote a book called *The Skeptical Environmentalist,* in which he argued that a statistical analysis of key global environmental indicators revealed that while there were environmental problems they were not as serious as was popularly believed.
Competitive Enterprise Institute, U.S.	This group focuses claims to "dispel the myths of global warming by exposing flawed economic, scientific and risk analysis."
James Annan, British climate scientist	He says the risks of extreme *climate sensitivity* and catastrophic consequences have been overstated.

Other skeptics state that global warming is not an environmental problem and that climate models that have been developed largely misrepresent reality. Others support the viewpoint that the data is misrepresented because most of the observations are taken in cities where temperatures are higher due to the urban heat island effect (cities are already warmer because asphalt and other dark manmade surfaces absorb enormous amounts of heat during the day). In reality, however, temperatures taken in urban areas are adjusted to compensate for that factor so the reading is unbiased. In addition, conservative politicians and a few scientists—many with ties to energy companies—claim global warming is insignificant or just a "manufactured crisis."

THE MIDDLE GROUND

According to a study in the *New York Times* from January 1, 2007, a new group has recently spoken out about global warming. This new outlook falls in the middle ground. These experts are challenging both extremes, instead looking at realities they believe may be somewhere closer to the middle.

Those favoring the middle ground support the idea that while the increasing accumulations of CO_2 in the atmosphere do pose a very real problem that does need to be dealt with, the methods used to deal with the problem need to be both logical and practical.

According to Carl Wunsch, who is a climate and ocean expert at the Massachusetts Institute of Technology, "It seems worth a very large premium to insure ourselves against the most catastrophic scenarios. Denying the risks seems utterly stupid. Claiming we can calculate the probabilities with any degree of skill seems equally stupid."

The following supporters in the middle ground believe the best solution is to reduce the overall vulnerability to all climate extremes, while simultaneously building public support for a shift to environmentally friendly energy sources. This group is not as willing to infer connections between specific events and global warming either. For example, they are much more conservative when they discuss the increasing damage in recent years due to specific weather incidences. In reference to recent damage done by hurricanes, their outlook is that as temperatures continue to rise, storms are likely to intensify, but there is no conclusive evidence of specific hurricanes being triggered specifically by global warming. Instead, they counter that the increased destruction from hurricanes is becoming more prevalent today because more people are building homes along the coast than ever before.

According to Dr. Roger A. Pielke, Jr., a political scientist, "We do have a problem, we do need to act, but what actions are practical and pragmatic?"

Dr. Mike Hulme, director of the Tyndall Center for Climate Change Research in Britain, believes that "Shrill voices crying doom could paralyze instead of inspire. I have found myself increasingly chastised by climate change campaigners when my public statements and lectures on climate change have not satisfied their thirst for environmental drama. I believe climate change is real, must be faced, and action taken. But the warning of catastrophe is in danger of tipping society onto a negative, depressive, and reactionary trajectory." He also stresses that it is important not to gloss over uncertainties; tackling the uncertainties is important so that issues do not get stretched out of proportion and misdiagnosed or misunderstood.

THE FAR LEFT—BELIEVERS IN GLOBAL WARMING

According to the UCS, during the 1990s climate change came to be accepted as "one of the biggest, most complex scientific and political challenges the world has ever faced. The issue is not amenable to simple solutions, and it is likely to be a pressing issue for the next century and beyond."

Climate change, or global warming, is indeed a serious issue in need of immediate action. Scientific research and discovery have progressed over the past few years to allow better understanding of many of the concepts involved in defining, measuring, and assessing global warming. Progress has been made in the marine and oceanographic sciences (ocean sediment cores, coral *bleaching,* acidification, current circulation, chemistry balance, and ice core analysis) and in geology and geomorphology (landforms and sediment analysis). Scientists worldwide have become involved, enabling the problem to be viewed as a global issue rather than a regional one.

The IPCC has also done an exceptional job of bringing the global scientific community together by building a team of more than 2,500 scientists all working toward a common goal. This effort has enabled the collection of standardized data.

The IPCC has also played a vital role in the political arena. To date, it has been very successful in informing both the general public and policy makers about sound science. The IPCC has also been able to organize the findings, facts, and results in a way that are meaningful and relevant to the public. According to the UCS, fewer skeptics are publicly challenging the existence of global warming or denying that global temperature has risen in this century in some part due to human activities. As more evidence of climate change appears and as climate modeling technology continues to advance and improve, support continues to grow in acknowledgement of the global warming problem and the necessity to take positive action to reduce it.

According to a *New York Times* editorial from November 20, 2007, when the IPCC released their latest report in 2007 on the status of global warming, "The scientists have done their job. Now it's time for the world leaders to do theirs."

Unlike IPCC's previous report issued in 2001, the latest report leaves absolutely no doubt that human-caused emissions from the burning of fossil fuels, methane from production of animals for food, and deforestation, among other causes, have been responsible for the steady rise in atmospheric temperatures. The report predicts, "If these emissions are not brought under control, the consequences could be disastrous—there would be further melting at the poles, sea levels rising high enough to submerge island nations, the elimination of one-fourth or more of the world's species, widespread famine in countries such as Africa, and more violent hurricanes."

The IPCC also warns that the problem needs immediate attention—humans do not have the luxury of time anymore. In fact, if greenhouse gases are not at least stabilized by 2015 and reduced immediately after so that all carbon-emitting technologies are gone by 2050, global warming will advance beyond control. The leader of the IPCC, Rajendra Pachauri, who is both a scientist and an economist, says, "What we do in the next two or three years will define our future."

Another *New York Times* article from February 3, 2007, reported that upon the IPCC release of their fourth report, scientists for the first time ever said that "global warming is 'unequivocal' and that human activity is the main driver, 'very likely' causing most of the rise in temperatures since 1950."

According to the IPCC, the Earth is in for centuries of rising temperatures, shifting weather patterns, rising seas, droughts, wildfires, and extinctions—results that will happen because of all the heat-trapping gases that are already in the atmosphere. Determined by each gas's *global warming potential* (GWP), there is a specific life span during which it traps heat. Carbon is the standard with a GWP of 1; all other gases are measured against it. Some gases may exist in small quantities but if their GWP is long, they can pose a serious problem.

Achim Steiner, executive director of the United Nations Environment Programme (UNEP) says, "The release of the IPCC's fourth report (released February 2, 2007) will be remembered as the date when uncertainty was removed as to whether human beings had anything to do with climate change on this planet. The evidence is on the table."

The difference in certainty levels changed in the IPCC's reports from 2001 to 2007. In the IPCC's 2001 report, their certainty level concerning global warming was likely, which means 66–90 percent certain. In their 2007 report, they categorized their prediction concerning global warming as very likely, meaning better than 90 percent, making the 2007 report much stronger.

According to John P. Holdren, an energy and climate expert at Harvard, "Since 2001, there has been a torrent of new scientific evidence on the magnitude, human origins and growing impacts of the climate changes that are under way. In overwhelming proportions this evidence has been in the direction of showing faster change, more danger, and greater confidence about the dominant role of fossil fuel burning and tropical deforestation in causing the changes that are being observed."

Dr. Richard B. Alley, a lead IPCC author and a professor at Pennsylvania State University, said about the 2007 report: "Policy makers paid us to do good science, and now we have very high scientific confidence in this work—this is real, this is real, this is real. So now act, the ball's back in your court."

Hervé Le Treut, an atmospheric physicist at Centre national de la recherche scientifique (CNRS) in France and lead author on the IPCC report, said, "By 2001, there were many signs that climate is changing and now we are already seeing the patterns that were described."

Kevin Trenberth, head of the climate analysis section at the U.S. National Center for Atmospheric Research (NCAR), said, "Northern Hemisphere snow cover has decreased and Arctic sea ice has been at record low levels in the past three years."

One of the aspects that believers in global warming focus on is the existence of several satellite temperature time series of the atmosphere. Thomas Peterson, a climate analyst for NOAA, said, "The connection of the satellite temperature collection error made the time series show more warming and is part of the reason why you no longer hear skeptics say that satellites don't show any warming."

In addition to the satellite data, massive amounts of other consistent data are being collected on the oceans via tide gauges and approximately 1,250 data-collecting buoys. This is seen as a real bonus because the world's oceans are the largest natural heat sinks. Not only are surface temperatures collected, but vertical profiling floats that record both temperature and salinity every 10 days to depths of 0.6–1.2 miles (1–2 km) are in place, according to Sydney Levitus, director of NOAA's World Data Center for Oceanography. Through these monitoring systems, NOAA has been able to determine that during the past 100 years sea level has risen at an average rate of 0.07 inch (0.17 cm) per year. They attribute the majority of this to *thermal* expansion of the top 2,297 feet (700 m) of the ocean water.

Kevin Trenberth at NCAR also said, "The human signal has clearly emerged from the noise of natural variability. Numerous changes in climate have been observed at the scales of continents or ocean basins. These include wind patterns, precipitation, ocean salinity, sea ice, ice sheets and aspects of extreme weather."

A new debate that has recently emerged from some believers in global warming is over the process of just how to slow it down. *A New York Times* study on April 17, 2008, pointed out that most of the focus up to now has been on imposing caps, or ceilings, on GHG emissions to encourage energy users to conserve energy and/or switch to renewable, nonpolluting energy resources. The IPCC and the 2008 presidential candidates have backed this market-based approach. Another group is now becoming heard. This group's (including economists, scientists, and students of energy policy) viewpoint is that using the cap approach is too little, too late.

The following table lists what some prominent believers in global warming are currently saying:

What Believers in Global Warming Are Saying

SOURCE	COMMENT/ACTIONS
Dave Stainforth, climate modeler at Oxford University	"This is something of a hot topic, but it comes down to what you think is a small chance—even if there's just a half percent chance of destruction of society, I would class that as a very big risk."
Dr. Rajendra Pachauri, chairman of the IPCC	He personally believes that the world has "already reached the level of dangerous concentrations of carbon dioxide in the atmosphere" and calls for immediate and very deep cuts in pollution if humanity is to survive.
Drew Shindell, NASA Goddard Institute for Space Studies	Believes global warming will cause serious drought in some areas. "There is evidence that rainfall patterns may already be changing. If the trend continues, the consequences may be severe in only a couple of decades."
James Overland, NOAA oceanographer	Believes that by 2050, the summer sea ice off Alaska's north coast will probably shrink to half of what it covered in the 1980s. This will have a profound effect on mammals dependent on the sea ice, such as polar bears, which could become extinct.
Ilsa B. Kuffner, USGS	Says oceans are becoming more acidic due to rising CO_2 in the atmosphere. This, in turn, is destroying the world's coral reefs.

SOURCE	COMMENT/ACTIONS
Shea Penland, a former coastal geologist at the University of New Orleans who died in 2008	Said the rate of sea-level rise has increased significantly over recent years and warns, "We're living on the verge of a coastal collapse."
World Wildlife Fund	One of their top priorities is to limit global warming and reduce emissions of CO_2. "If we want to have something left to protect at all, the managers of protected areas need to assess the climate change impacts and prepare their parks for the worst."
James E. Hansen, director, NASA GISS	"As we predicted last year, 2007 was warmer than 2006, continuing the strong warming trend of the past 30 years that has been confidently attributed to the effect of increasing human-made greenhouse gases."
Terrence Joyce, Woods Hole physical oceanography department	Concerning changes to the Ocean Conveyor Belt: "It could happen in 10 years. Once it does, it can take hundreds of years to reverse." He is alarmed that Americans have yet to take the threat seriously. In a letter to the *New York Times* last April, he wrote, "Recall the coldest winters in the Northeast, like those of 1936 and 1978, and then imagine recurring winters that are even colder, and you'll have an idea of what this would be like."

These are merely a few examples in a sea of controversial opinions. Each individual ultimately must form their own opinion. The best way to do that is to study the facts, become aware of the issues, and pay attention to what is happening not only to the local environment but also to the global one. In the end, the judgments are personal, as are any actions associated with them.

Green Energy and Global Warming Research

Today, climatologists can study climate patterns using sophisticated models of the Earth's atmosphere and oceans. As technology has advanced and mathematical models have become more sophisticated, through the process of matching observed and modeled patterns, they have been able to tease out the human fingerprints that are associated with the changes, further solidifying the proof that humans are having an impact on the environment, and climate in particular. This chapter discusses renewable energy sources and their role in managing global warming, as well as current research that scientists are involved in as they strive to find solutions to the problem and provide sustainable, environmentally friendly energy for the future.

THE ENVIRONMENTAL BENEFITS OF GREEN ENERGY

As the world's environmental consciousness has risen over the past two decades from paying attention to oil spills, chemical spills and leaks, nuclear reactor accidents, overwhelming pollution, massive deforestation of the world's rain forests, ozone depletion, and global warming, people have focused on alternate sources of energy, environmentally friendly, sustainable, renewable energy. Also added to the mix have been oil embargoes, shortages, and skyrocketing oil prices. The combination of environmental damage, public education from the world's leading environmental organizations, and an uncertain economy with unstable oil prices have begun to fuel a wave of concern for the health of the environment and the future of life on Earth.

Green energy—also referred to as renewable energy—includes solar power, wind power, geothermal power, hydropower, biofuels, and ocean energy. Renewable energy is also referred to as sustainable energy—the generation of energy that meets the needs of society today without compromising the ability of future generations to be able to produce the necessary energy to meet their own needs. Unlike fossil fuels, the finite supplies of which are only expected to last another 170 years if the present rate of consumption continues, renewable energy sources can be produced indefinitely, without harming the environment or adding to global warming.

The current average global growth rate of energy use is 1.7 percent. If that rate continues, then by 2030, the amount of energy consumed will double, compared to the amount of energy used in 1995; by 2060, it will have tripled. Increasing demands for energy pose serious environmental and health problems for future generations—especially through global warming. Right now, the current production and use of energy causes more damage to the environment than any other single human activity—it contributes to 80 percent of the air pollution suffered by major cities worldwide and more than 88 percent of the greenhouse gas emissions responsible for global warming.

Currently, many areas in the United States offer green energy, such as renewable sources of electricity. Many states have green pricing programs offered by their local utilities. A green pricing program is a voluntary utility-sponsored program that enables customers to support the development of renewable resources. Participating customers may pay a premium on their electric bill to cover the incremental cost

A layer of smog appearing over Denver, Colorado *(David Parsons. DOE/NREL)*

of the renewable energy. When customers purchase green electricity, they ensure that the power provider will add that amount of renewable power into the grid, offsetting the need for the same amount of conventional power (power produced through the burning of fossil fuels, usually coal).

Back in 2003, the Bush administration included funding for renewable energy programs, seeking $555 million in clean energy tax incentives as the first part of a $4.6 billion commitment over the next five years. The tax credits from these served to promote investments in renewable energy (solar, wind, and biomass), hybrid and fuel cell vehicles, cogeneration, and landfill gas conversion. The plan also tasked the U.S. Bureau of Land Management (BLM) with launching a major effort to increase its renewable energy activities by encouraging the research, exploration, and development of renewable energy resources from public lands. The U.S. Department of Energy (DOE) also supports renewable energy with plans to purchase 3 percent of its electricity from non-hydro renewable energy sources by 2005 and 7.5 percent by 2010. The U.S. Environmental Protection Agency (EPA) has identified the following benefits of purchasing green power:

- raises public awareness of renewable energy
- promotes the development of new renewable energy resources
- creates jobs in the renewable energy industry
- reduces emissions of greenhouse gases and air pollution

Many states have realized the benefits and cost savings of supporting green energy purchases for government facilities. Currently, Arizona, Colorado, Maryland, Nebraska, and New York have executive orders or legislation requiring state agencies to obtain a portion of their electricity needs from renewable energy sources. In addition to avoiding the depletion of natural resources, there are many benefits to using renewable energy sources, including environmental, economic, energy security, and employment. When a utility company uses renewable sources, it has a direct economic benefit for the company because it reduces their Clean Air Act compliance costs. When they do not have to invest in equipment necessary to reduce the emission of pollutants directly into the atmosphere, operating costs are cut.

One of the major external economic benefits of renewable energy is in the category of human health care costs, specifically in the form of reduced health treatment costs, lower health insurance rates, less missed work, and lower death rates. According to a survey of health impacts conducted by the Pace University School of Legal Studies and studies conducted by the American Lung Association, the annual U.S. health costs from all air pollutants may be as high as hundreds of billions of dollars. Both industry and individuals will gain by using renewable energy sources because these sources produce very little or no pollution. Environmental regulations usually focus on one pollutant at a time as scientific research is conducted on that pollutant and regulations are put in place based upon that research. When the government imposes a new regulation, industry may add a series of new pollution controls. The problem that sometimes occurs is that if further research is done and another pollutant is discovered, researched, and a control put in place for that pollutant, industry will have to then adapt to that new pollutant, often having to add additional costly equipment to control the new pollutant's emission. As this adds up, it can be very expensive for industry, which then spreads the cost to the customer. The Union of Concerned Scientists (UCS) argues that replacing fossil fuel generators with renewable energy technology may seem expensive at first, but if future controls are considered, renewable technology may indeed be the most cost-effective alternative right from the beginning.

As an example, UCS used the following situation. In 1998, there were many new environmental regulations pending, which would directly affect industries powered by fossil fuels.

- The level of ozone (smog) allowed in ambient air was being reduced from 0.18 to 0.08 *parts per million* (ppm).
- Nitrogen oxides were under new consideration under the Clean Air Act.
- Sulfur dioxide limits were to be tightened in the year 2000 when Phase II of the Clean Air Act went into effect. This would affect every coal-burning power plant in the United States.
- Fine particles were being regulated for the first time, with final rules put in place by 2005.
- Mercury and other toxic metals have been the subject of substantial research by the EPA. The EPA announced it would require coal-fired plants to disclose discharges, and it would use the data to decide on regulations by 2000.
- Carbon dioxide (CO_2) emissions would need to be reduced to implement the Kyoto agreement on global warming.

Now, with all of these regulations pending, conversion to renewable technologies in 1998 would have forestalled the need for expensive future retrofits to achieve compliance with these regulations. Therefore, there is much to be gained economically (and saved) by making investments in renewable energy.

A study conducted in 1997 entitled "The Hidden Benefits of Climate Policy: Reducing Fossil Fuel Use Saves Lives Now" illustrates the benefit of multi-emission reductions. The study concluded that measures to reduce global carbon dioxide emissions—including increasing the use of renewables—could save 700,000 lives each year and a cumulative total of 8 million lives worldwide by 2020.

According to the UCS, there are also diversity and energy security benefits. Renewables add an economically stable source of energy for the United States. When a country depends on only a few types of energy sources—such as coal, oil, and gas—it puts them in a much more vulnerable position, held ransom to factors such as political unrest, volatile prices, and interruptions in fuel supplies. This has played out historically during the oil embargo that occurred in 1973 and in 2008 when OPEC raised the price of a barrel of oil over $100. Because most renewable forms of energy do not depend on fuel markets, they are not subject to price fluctuations resulting from increased demand, decreased supply, or manipulation of the market.

There are also economic development benefits. Renewable energy technologies keep money in the United States and create significant regional benefits through economic development. Renewable technologies create jobs using local resources in new, green, high-tech industries with an enormous export potential that is just waiting to be tapped. They also create jobs in local industries, such as banks and construction firms. In fact, during the 1990s, the U.S. renewable electricity industry employed more than 117,000 people.

According to the UCS, renewables create increased revenues for local landowners. For instance, farmers can increase their returns on their land by 30 to 100 percent if they lease part of it for wind turbines while continuing to farm it. Another study conducted by UCS found that adding 10,000 MW of wind capacity nationally would generate $17 million per year in land-use easement payments to the owners of the land on which the wind farms are situated and $89 million per year from maintenance and operations.

Renewables are also significant income resources for local tax bases. For instance, wind farms in California currently pay $10 to $13 million in property taxes. According to the American Wind Energy

Association, at least 44 states are involved in manufacturing wind energy system components.

The UCS reports that an analysis they completed for Wisconsin found that, over a 30-year period, an 800-megawatt mix of new renewables would create about 22,000 more jobs per year than new natural gas and coal plants would. A New York State energy office study concluded that wind energy would create 27 percent more jobs than coal and 66 percent more than a natural gas plant. Economic Research Associates completed a study of energy efficiency and renewable energy as an economic development strategy in Colorado and found an energy bill savings of $1.2 billion for Colorado ratepayers by 2010 with a net gain of 8,400 jobs. The California Energy Commission estimates that the new renewable industries that will be built using $162 million in public funding will bring in:

- $700 million in private capital investment
- 10,000 construction jobs, with over $400 million in wages
- 900 ongoing operations and maintenance jobs with $30 million in long-term salaries
- gross state product impacts of $1.5 billion during construction and $130 million in annual ongoing operations

As an additional bonus, in addition to creating jobs, renewables can improve the economic competitiveness of a region by enabling it to avoid additional costly environmental controls on other industries, as well as by keeping long-term energy prices stable. Renewables also contribute to economic development by providing opportunities to build export industries. In developing countries that do not have electricity grids, pipelines, or other energy infrastructure already in place, renewable energy technologies can be a cost-effective solution in providing these areas with electricity, such as rural areas. The American Wind Energy Association has estimated that global markets for wind turbines alone will amount to as much as $400 billion between 1998 and 2020.

The United States is not the front-runner in promoting renewable energy resources. Japan and various European nations are in the lead globally by already encouraging the development of renewables through providing greater subsidies than the United States currently does. The United States is currently in a position to learn by the examples of several foreign countries that already understand the importance of conservation and environmental protection. For years, other countries have not had access to inexpensive fuels for their cars and homes and have had to adjust accordingly. The United States is in a position now where they have an opportunity to learn from their neighbors—and must use that opportunity—about fuel efficiency and sustainable energy practices if the problem of global warming is to be successfully addressed. One major lesson to be learned is that by increasing renewables, there are many associated benefits.

Prior to the 1980s, the only widely used renewable electricity technology used in the United States was hydropower. It is still the most significant source of renewable energy, producing 20 percent of the world's electricity and 10 percent of that of the United States. The 1973 oil crisis grabbed the nation's attention as to its vulnerability because of its dependence on foreign oil. It was the resulting subsequent changes in federal policy that spurred the development of renewable technologies other than hydro.

In 1978, Congress passed the Public Utility Regulatory Policies Act (PURPA), which required utilities to purchase electricity from renewable generators and from cogenerators (which produce combined heat and power, usually natural gas) when it was less expensive than electric utilities could generate themselves. Some states—especially California and those in the Northeast—required utilities to sign contracts for renewables whenever electricity from those sources was expected to be cheaper over the long term than electricity from traditional sources. It was these states that had the largest growth of renewables development under PURPA. However, because oil price projections were high and because utilities were planning expensive nuclear plants at the time, these renewables contracts turned out to be expensive relative to the low fossil fuel prices of the 1990s, striking a heavy blow to the program.

Even so, under PURPA over 12,000 megawatts of non-hydro renewable generation capacity came online, which enabled renewable technologies to develop commercially. Wind turbine costs, for instance, decreased by more than 80 percent. Over the past five years, renewable energy growth has been modest, averaging less than 2 percent per year, primarily because of the low cost of fossil fuels. In addition, the uncertainty around the deregulation of the utility industry served to freeze investments in renewables, as utilities avoided new long-term investments.

Current levels of renewables development represent only a tiny fraction of what could be developed. Many regions of the world and the United States are rich in renewable resources. Winds in the United States contain energy equivalent to 40 times the amount of energy the nation uses. The total sunlight falling on the nation is equivalent to 500 times America's energy demand. Accessible geothermal energy adds up to 15,000 times the national demand. There are, however, limits to how much of this potential can be used, because of competing *land uses,* competing costs from other energy sources, and limits to the transmission system needed to bring energy to end users. Solar, geothermal, wind, hydropower, biofuels, and ocean energy are the renewables that are being looked to to supply the energy of the future.

SOLAR ENERGY

Solar energy can be used directly as an energy source to generate heat, lighting, and electricity. The amount of energy from the Sun received by the Earth's surface each day is enormous. As a comparison, all of the energy currently stored in the Earth's reserves of coal, oil, and natural gas is roughly equivalent to 20 days of the solar energy that reaches the Earth's surface.

Outside the Earth's protective atmosphere, the Sun's energy contains roughly 1,300 watts per square meter. Approximately one-third of this light is reflected back into space, and some is absorbed by the Earth's atmosphere. When the solar energy finally reaches the Earth's surface, the energy is roughly equivalent to about 1,000 watts per square meter at noon on a cloudless day. According to the UCS, when this is averaged over the entire surface of the planet, 24 hours a day for an entire year, each square meter collects the energy equivalent of almost a barrel of oil each year, or 4.2 kilowatt-hours of energy every day.

As shown in the figure, geographic areas vary in the amount of storable, usable energy they receive. Deserts with very dry, hot air and minimal cloud cover (such as the southwestern United States) receive

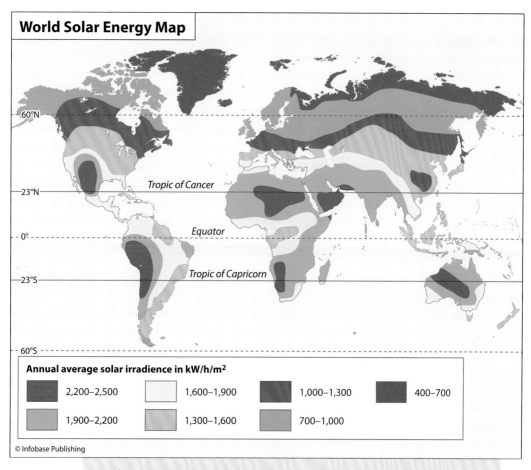

A solar resource map of the world—the more solar energy that is received, the greater the potential is to use solar power as a sustainable energy source.

the most sun (more than six kilowatt-hours per day per square meter). Northern climates (such as the northeastern United States) receive less energy (about 3.6 kilowatt-hours). Sunlight also varies by season, with some areas receiving very little sunshine during the winter due to extremely low sun angles. Seattle in December, for example, only gets about 0.7 kilowatt-hours per day.

Solar collectors used to capture solar energy do not capture the maximum available solar energy. Depending on the collector's efficiency, only a portion of it is captured. One method of using solar energy is

132   **CLIMATE MANAGEMENT**

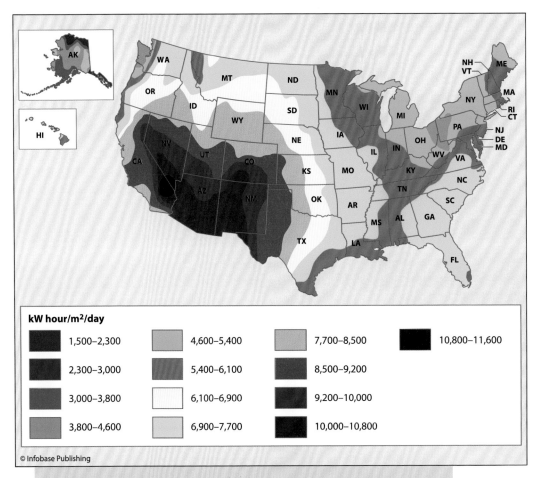

A solar resource map of the United States

through passive collection in buildings—designing buildings to use natural sunlight. Passive solar energy refers to a resource that can be tapped without mechanical means to help heat, cool, or light a building. If buildings are designed properly, they can capture the Sun's heat in the winter and minimize it in the summer, using natural daylight all year long. South-facing windows, skylights, awnings, and shade trees are all techniques for exploiting passive solar energy.

According to studies conducted by the UCS, residential and commercial buildings account for more than one-third of U.S. energy use. Solar design, better insulation, and more efficient appliances could reduce the demand by 60 to 80 percent. New construction can employ specific design features, such as orienting the house toward the south, putting most of the windows on the south side of the building, and taking advantage of cooling breezes in the summer. These are inexpensive and effective ways to make a home more comfortable and efficient, thereby reducing its global warming potential (from decreased fossil fuel use because electricity or natural gas did not have to be used to artificially heat or cool the home). Today, several hundred thousand passive solar homes exist in the United States.

In addition to passive systems, there are also active systems. These systems actively gather and store solar energy. Solar collectors are often placed on rooftops of buildings to collect solar energy. The energy can then be used for space heating, water heating, and space cooling. These collectors are usually large, flat boxes painted black on the inside and covered with glass. Inside the box, pipes carry liquids that transfer the heat from the box into the building. The heated liquid (usually a water/alcohol mixture to prevent freezing) is used to heat water in a tank or is passed through radiators that heat the air.

Based on data collected by the UCS, currently about 1.5 million U.S. homes and businesses use solar water heaters (less than 1 percent of the U.S. population). Solar collectors are much more common in other countries. In Israel, for example, they require that all new homes and apartments use solar water heating. In Cyprus, 92 percent of the homes already have solar water heaters. The UCS believes that the number of solar water heaters and space heaters in the United States may rise dramatically in the next few years due to the skyrocketing prices of natural gas.

According to the DOE, water heating accounts for 15 percent of an average household's energy use. As the price rates for natural gas and electricity continue to climb as they have recently, it will continue to cost more to heat water supplies. The DOE predicts that in the near future, more homes and businesses will start heating their water supplies through solar collectors. Using solar energy could save homeowners between $250 and $500 per year depending on the type of system being replaced.

Solar energy can also be generated through solar thermal concentrating systems. These systems use mirrors and lenses to concentrate the rays of the Sun and can subsequently produce extremely high temperatures—up to 5,432°F (3,000°C). This intense heat can also be used in industrial applications to produce electricity.

Solar concentrators come in three designs: parabolic troughs, parabolic dishes, and central receivers. The most commonly used are the parabolic troughs. These have long, curved mirrors that concentrate sunlight on a liquid inside a tube that runs parallel to the mirror. The liquid is heated to about 572°F (300°C) and runs to a central collector, where it produces steam that drives an electric turbine. Parabolic dish concentrators are similar to trough concentrators but focus the sunlight onto a single point. Dishes can produce even higher temperatures, but these systems are much more complicated, need more development, and therefore, are not used much at this point. The third type is a central receiver. These systems employ a power tower design, where a huge area of mirrors concentrates sunlight on the top of a centralized tower. The intense heat boils water, producing steam that drives a 10-megawatt generator at the base of the tower.

Presently, the parabolic trough has the greatest commercial success, mainly due to the nine solar electric generating stations (SEGS) that were built in California's Mojave Desert from 1985 to 1991. These stations range in capacity from 14 to 80 megawatts, with a total capacity of 354 megawatts. Each plant is still in operation.

Due to several state and federal policies and incentives, more commercial-scale solar concentrator projects are under development. Currently, modified versions of the SEGS plants are being constructed in Arizona (1 megawatt) and Nevada (65 megawatts). In addition, Stirling Energy Systems began building a 500-megawatt facility in California's Mojave Desert in 2005 using a parabolic dish design with plans to become operational in 2009 in order to supply power to Southern California under a 20-year contract to meet the requirements in the state's renewable electricity standard.

Solar cells—or photovoltaics (PV)—are another key form of solar energy. In 1839, the French scientist Edmund Becquerel discovered that certain materials gave off a spark of electricity when struck with sunlight. This photoelectric effect was demonstrated in primitive solar cells constructed of selenium in the late 1800s. Later, in the 1950s, scientists

Stretched membrane heliostats with silvered polymer reflectors will be used as demonstration units at the Solar Two central receiver in Daggett, California. The Solar Two project will refurbish this 10-megawatt central receiver power tower known as Solar One. *(Sandia National Laboratories. DOE/NREL)*

at Bell Labs used silicon and produced solar cells that could convert 4 percent of sunlight energy directly into electricity. Within a few years, these photovoltaic cells were powering spaceships and satellites.

The most critical components of a PV cell are the two layers of semiconductor material that are composed of silicon crystals. Boron is added (to make the cell more conductive) to the bottom layer of the PV, which bonds to the silicon and creates a positive charge. Phosphorus is added to the top to make it more conductive and to produce a negative charge.

An electric field is produced that only allows electrons to flow from the positive to the negative layer. Where sunlight enters the cell, its energy knocks electrons loose on both layers. The electrons want to flow from the negative to positive layer, but the electric field prevents this from happening. The presence of an external circuit, however, does provide the necessary path for electrons in the negative layer to travel to the positive layer. Thin wires running along the top of the negative layer provide an external circuit, and the electrons flowing through this circuit provide a supply of electricity.

Most PV systems consist of individual cells about four inches (10 cm) square. Alone, each cell generates very little energy—less than two watts; so they are often grouped together in modules. Modules can then be grouped into larger panels encased in glass or plastic to provide protection from the weather. Panels can further be grouped into even larger arrays. The three basic types of solar cells made from silicon are single-crystal, polycrystalline, and amorphous.

Since the 1970s, serious efforts have been underway to produce PV panels that can provide cheaper solar power. Innovative processes and designs are constantly being released on the market and driving prices down. These include inventions such as photovoltaic roof tiles and windows with a translucent film of amorphous silicon (a-Si). The growing global PV market is also helping reduce costs.

In the past, most PV panels have been used for off-grid purposes, powering homes in remote locations, cellular phone transmitters, road signs, water pumps, and millions of solar watches and calculators. The world's developing nations look at PV as a viable alternative to having to build long, expensive power lines to remote areas. In the past few years, in light of global warming and rising energy costs, the PV industry has been focused more on homes, businesses, and utility-scale systems that are actually attached to *power grids.*

In some areas, it is less expensive for utilities to install solar panels than to upgrade the transmission and distribution system to meet new electricity demand. In 2005, for the first time, the installation of PV systems connected to the electric grid outpaced off-grid PV systems in the United States. According to the DOE, as the PV market continues to expand, the demand for grid-connected PV will continue to climb. The

NEW WAYS TO STORE SOLAR ENERGY

According to a *New York Times* report on April 15, 2008, solar power has always faced the problematic issue of how to store its energy so that the demand for electricity can be met at any time—even at night or when the Sun is not shining. In the past, this has been a problem because electricity is difficult to store and batteries cannot efficiently store energy on a large scale. The solar power industry is now trying a new approach—the concept of capturing the Sun's heat.

The idea, according to John S. O'Donnell of Ausra, a solar thermal business, is that heat can now be captured and stored cost-effectively and "That's why solar thermal is going to be the dominant form [of solar energy]." In the concept he is referring to, solar thermal systems are built to gather heat from the Sun, boil water into steam, spin a turbine, and generate power—just as present-day solar thermal power plants do—but not immediately. Instead, the heat would be stored for hours, or even days, like the water holding energy behind a dam. In this way, a power plant could store its output and could then pick the time to sell the production based on need, expected price, or whatever criteria it deemed. In this way, energy could be realistically promised even if the weather forecast was unfavorable or uncertain.

Another solar energy company has the same goals but approaches it a bit differently. They use a power tower, which is like a water tank on stilts surrounded by hundreds of mirrors that tilt on two axes—one to follow the Sun across the sky during the course of the day and the other in the course of the year. In the tower and in a tank below, there are tens of thousands of gallons of molten salt that can be heated to very high temperatures but not reach high pressure. According to Terry Murphy, the president and chief executive of Solar Reserve, "You take the energy the Sun is putting into the Earth that day, store it and capture it, put it into the reservoir, and use it on demand." In Murphy's design, his power tower will supply 540 megawatts of heat. At the high temperatures it could achieve, that would produce 250 megawatts of electricity—enough to run an average-sized city.

"It might make more sense to produce a smaller quantity and run well into the evening or around the clock or for several days when it is cloudy," Murphy said.

(continues)

(continued)

The tower design can also be operated at higher latitudes and places with less Sun. The array would just have to be built with bigger mirrors. Interestingly, Murphy helped construct a power tower at a plant in Barstow, California, in the late 1990s that worked well. Then the price of natural gas dropped, and the plant turned to that fuel source instead to power the plant. Murphy's response was, "There were no renewable portfolio standards. Nobody cared about global warming, and we weren't killing people in Iraq."

UCS believes that solar energy technologies will face significant growth during the 21st century because of new knowledge about global warming. By 2025, the solar PV industry aims to provide half of all new U.S. electricity generation.

Aggressive financial incentives in both Germany and Japan have made them world leaders in solar energy use. The United States is just now beginning to pick up momentum. In January 2006, the California Public Utility Commission approved the California Solar Initiative, which dedicates $3.2 billion over 11 years to develop 3,000 megawatts of new solar electricity. This is the equivalent of placing PV systems on 1 million rooftops. Other states are now following California's lead. New Jersey, Colorado, Pennsylvania, and Arizona all have specific requirements for solar energy written into plans as part of their renewable electricity standards. Other states are now offering rebates, production incentives, tax incentives, and loan and grant programs.

The federal government, in trying to promote renewable energy, is also offering a 30 percent tax credit (up to $2,000) for the purchase and installation of residential PV systems and solar water heaters. As the population increasingly shifts to solar energy, it plays an integral role in ending the nation's dependence on foreign sources of fossil fuels, further combats global warming, and promotes a more secure future based on clean, sustainable energy.

GEOTHERMAL ENERGY

Geothermal energy involves the latent heat of the Earth's core. Geothermal resources are not new; they have been used for centuries—natural hot springs have been used worldwide for cooking, bathing, and heating bathhouses. In 1904, inhabitants in Tuscany, Italy, were the first to actually generate electricity from geothermal water. Geothermal energy exists naturally in several forms, such as:

- In hydrothermal reservoirs of steam or hot water trapped in rock. These reservoirs are found in specific regions and are the result of geologic processes.
- In the heat of the shallow ground. This Earth energy occurs everywhere and is the normal temperature of the ground at shallow depths. Specific geologic processes do not enhance it, so it is not as hot as other geothermal sources.
- In the hot, dry rock found everywhere between five and 10 miles (8–16 km) beneath the Earth's surface and at even shallower depths in areas of geologic activity.
- In magma, molten or partially molten rock that can reach temperatures of up to 2,192°F (1,200°C). Some magma is found at shallower depths, but most is too deep beneath the Earth's surface to be reached by current technology.
- In geopressurized brines. These are hot, pressurized waters containing dissolved methane that are found 10,000–20,000 feet (3,048–6,096 m) below the surface.

With current technology, only hydrothermal reservoirs and Earth energy sources supply geothermal energy on a large scale. Hydrothermal reservoirs are tapped by existing well drilling and energy-conversion technologies to generate electricity or to produce hot water for direct use. Earth energy is converted for use by geothermal heat pumps.

In order to be useful, a carrier fluid such as water or gas must convey the heat. In hydrothermal reservoirs, the fluid is found naturally

Geothermal power plant at The Geysers near Calistoga, California
(Lewis Stewart, DOE/NREL)

in the form of groundwater. A carrier fluid can be artificially added to create a geothermal system. Geothermal heat pumps, for example, that use Earth energy sources to provide heating and cooling for buildings circulate a water or antifreeze solution through plastic tubes. This solution removes heat from, or transfers heat to, the ground. There is never any contact between the fluid, groundwater, or Earth.

The temperature of the carrier fluid determines how the geothermal energy can be used. The hotter the fluid, the more applications there are. Thermal fluids that are at the steam phase—temperatures above 212°F (100°C)—can be used for industrial-scale *evaporation* such as drying timber. Lower temperature thermal heat—less than 212°F (100°C)—in the form of hot water can be used to heat homes, power district heating systems, or for small-scale evaporation processes such as food drying.

Geothermal heat pumps that use Earth energy sources to supply direct heat to homes are the most efficient technology available for heating and cooling, producing three to four times more energy than they consume. They can reduce the peak generating capacity for residential installations by 1–5 kW and can be used effectively even with a wide range of ground temperatures. The successful generation of electricity usually requires higher temperature fluids—above 284°F (140°C). Geothermal power plants use wells to draw water from depths of 0.6–1.9 miles (1–3 km) and produce electricity in one of two types of plants: steam turbine plants or binary plants.

Steam turbine plants release the pressure on the water at the surface of the well in a flash tank where some of the water "flashes" or explosively boils to steam. The steam then turns a turbine engine, which drives a generator to produce electricity. The water that does not boil to steam is injected back into the ground to maintain the pressure of the reservoir.

In a binary plant, instead of being flashed to steam, the water heats a secondary working fluid such as isobutene or isopentane through a heat exchanger. This secondary fluid is then vaporized and sent through a turbine to turn a generator after which it is cooled and *condensed* into a liquid again. It then travels back through the heat exchanger to be vaporized again. The water is injected back into the reservoir to recharge the system. Because the working fluids vaporize at lower temperatures than water, binary plants can produce electricity from lower temperature geothermal resources.

Globally, geothermal power plants supply approximately 8,000 MW of electricity and are used in many countries, including Italy, Japan, Iceland, China, New Zealand, Mexico, Kenya, Costa Rica, Romania, Russia, the Philippines, Turkey, El Salvador, Indonesia, and the United States. One of the major advantages of geothermal power plants is that they can remain online nearly continuously, making them much more reliable than coal-based power plants, which statistically are online and operational roughly 75 percent of the time. Geothermal systems can also be installed modularly, increasing power levels incrementally to fit current demand. They also use only a small amount of land in comparison to other types of power plants. In addition, that same land can be used simultaneously for other purposes, such as agriculture, with little interference or chance of an accident occurring. As an example, the Imperial Valley of Southern California, which is one of the most productive agricultural areas in the United States, also supports 15 geothermal plants that currently produce 400 MW of electrical power.

Geothermal energy is also viewed as an environmentally friendly energy resource. Geothermal power plants have very low emissions of sulfur oxide and nitrogen oxide (that cause acid rain) and CO_2 contributing to global warming. The typical lifetime for geothermal activity around magmatic centers is from 5,000 to 1 million years; a time interval so long that geothermal energy is considered to be a renewable resource. Although geothermal energy is site specific, it is viewed as a major renewable clean-energy resource, able to provide significant amounts of energy for today's energy demands.

WIND ENERGY

Wind is simply thermal power that has already been converted to mechanical power. As the wind turns the blades of a turbine, the rotating motion drives a generator and produces electricity without any emissions. The resultant wind power, or wind energy, can be employed for various tasks—it can pump water or be converted to electricity (through a turbine).

Modern wind turbines fall into two different groups: the horizontal-axis variety, like the traditional farm windmills used for pumping water, and the vertical-axis design, the eggbeater style. Wind turbines are often grouped together into a single wind power plant—also referred to as a wind farm—in order to generate bulk electrical power. Once electricity is generated from the turbines, it is fed into the local utility grid and distributed to customers just as it is with conventional power plants.

All electric-generating wind turbines, no matter what size, are comprised of the same basic components: the rotor (the piece that actually rotates in the wind), the electrical generator, a speed control system, and a tower. There are multiple sizes of turbines and lengths of blades, and each has its unique energy capacity, which can vary from several kilowatts to several megawatts, depending on the turbine design and the length of the blades. Most turbines produce about 600 kW, but more powerful machines are becoming more common as the market expands and technology improves. There are currently several different types of turbines available—with one, two, or three blades, different blade designs, and varying orientations to the wind. There are machines that have propeller blades that span more than the entire length of a football

Maple Ridge Wind Farm in Lewis County, New York *(IBERDROLA RE-NEWABLES, Inc., DOE/NREL)*

field—equivalent to a 20-story building in height—and produce enough electricity to power 1,400 homes. A small home-sized individual wind machine has rotors between eight and 25 feet (2.4–7.6 m) in diameter and stands 30 feet (9 m) tall and can supply the power needs of an all-electric home or small business.

With wind energy, geographic location is critical. Wind turbines cannot just be placed anywhere. They must be placed in areas where wind is not only available consistently, but the wind must also be able to maintain a certain wind speed. Wind speed is critical—the energy in wind is proportional to the cube of the wind speed. This means that a stronger wind provides much more power.

As far as new sources of electricity generation, wind energy has been the fastest growing. Worldwide, in the 1990s, wind energy use has grown at a rate of about 26 percent per year. It is also the most economically competitive energy of the renewable sources. The majority of the growth in the market has taken place in Denmark and Germany, because their government policies, coupled with high conventional energy costs, have made wind energy very attractive to residents of these countries. India has also experienced growth in the wind energy industry recently.

In the United States, the state that uses the most wind energy is California. The global wind energy industry has grown steadily over the past 10 years, and companies are beginning to compete. As the industry expands, new developments and improvements are taking place. A full range of highly reliable and efficient wind turbines is being developed. These new-generation turbines are able to perform at 98 percent reliability in the field, representing significant progress since the technology was first introduced as a sustainable energy resource in the early 1980s.

Even though wind is an intermittent source of power, unlike hydropower, wind energy is usually readily available at times of highest electricity demand. One major advantage to wind power technology is that turbines can be used as a single stand-alone unit in small groups to provide power locally, or they can be part of an energy system, either with other renewable energy sources or connected to the power grid.

As far as economics, there are currently several factors bearing on the cost of wind power, which affect its feasibility as a commercial energy source. The wind speed, the reliability and efficiency of the turbines, and the estimated rates of return on investment all determine what the cost of wind energy will be. Fortunately, with improved technology and manufacturing procedures, the cost of generating electricity from wind power has dropped to less than seven cents per kilowatt-hour, compared to four to six cents per kilowatt-hour to operate a new coal or natural gas power plant—and the process is expected to get even cheaper over the next 10 years.

Currently, new utility-scale wind projects are being built throughout the United States. Associated energy costs are ranging from 3.9 cents per kilowatt-hour (at very windy sites in Texas) to five cents or more (in the Pacific Northwest). According to the DOE, today in the United States wind energy provides more jobs per dollar invested than any other energy technology—currently calculated at more than five times that from coal or nuclear power. This technology uses expertise of several scientific fields such as engineering, electronics, aerodynamics, and materials sciences, creating a viable job market in those fields.

Another concept associated with wind energy that is becoming significant in the United States is that of net metering, which many states are now permitting. This is the concept in which the utility must buy wind power generated by homeowners at the same retail rate the utility charges. This essentially allows customers' meters to turn backward while wind energy is supplied to the grid by their turbines.

Wind energy is also significant in terms of global warming prevention. The amount of emissions avoided because of California's wind power plants in 1990 alone was more than 2.5 billion pounds of CO_2, and 15 million pounds of other pollutants. As a comparison, it would take a forest of 90 to 175 million trees to provide the same air quality.

One of the persistent downsides of this form of energy, however, is that even in spite of the significant decreases in costs over the past decade the technology still requires a higher initial investment than fossil-fueled generators. Of this, about 80 percent of the cost is the machinery, with the rest being the site preparation and installation. The minimal operating expenses and zero fuel bill offsets the high initial costs, but it is still difficult presently for some consumers to see the broader picture and the inherent benefits of choosing wind energy over fossil fuel energy.

Some critics claim there are some negative impacts to wind energy. Although these plants have relatively little impact on the environment, there is some concern over the noise produced by the rotor blades, the aesthetics, and occasional avian mortality (birds flying into the blades). Most of the problems have been significantly reduced through technological development or by properly situating wind plants, although avian mortality still remains an issue.

The major drawback to wind energy is that it is not a constant, dependable source of energy. There may be times when there is not enough wind blowing. This challenge can be overcome by using batteries. Also, good wind sites are often located in remote locations far from areas of electric power demands, such as in cities. In some places, wind resource development may compete with other uses for the land and those alternative uses may be more highly valued than electricity generation. On a positive note, wind turbines can be located on land that is also used for grazing or even farming.

The following lists the benefits of using wind energy, as designated by the EPA:

- reduced emissions of greenhouse gases, air pollutants, and hazardous wastes
- reduced reliance on imported energy
- no risk of fuel price hikes
- increased local job and business opportunities
- quick construction with options to build in phases according to need
- contribution to the local economy through the payment of property taxes and land rents

HYDROPOWER

Hydropower uses the energy of the hydrologic cycle, which is ultimately driven by the Sun, making it an indirect form of solar energy. Energy contained in sunlight evaporates water from the ocean and deposits it on land in the form of rain, snow, and other forms of precipitation. Precipitation that is not absorbed by the ground runs off the land into the ocean via the world's vast network of rivers and repeats the process. Hydroelectric plants built along rivers generate power by releasing water stored behind concrete dams built across the river to turn water turbines. The power plants capture the energy released by water falling through a turbine, which converts the water's energy into mechanical power. The mechanical energy of the rotating turbines drives generators to produce electricity.

Hydro dams are present in almost all regions of the world and have played a key role in development for thousands of years. Many modern dams are multipurpose, built primarily for irrigation, water supply, flood control, electric power, and improvement of navigation. They also provide recreation such as fishing, boating, water skiing, and swimming and become refuges for fish and birds. In the last two centuries, they have also played a key role in producing large-scale power and electric-

Ice Harbor Dam near Burbank, Washington. Hydroelectric power is a clean, renewable source of energy and generates about 10 percent of the energy in the United States. *(U.S. Army Corps of Engineers, DOE/NREL)*

ity. Dams also slow down streams and rivers so that the water does not carry away soil, thereby preventing erosion.

Hydroelectric power plants exist in many sizes from less than 100 kilowatts to several thousand megawatts. There are already more than 35,000 large dams in existence worldwide. The number and size of recent large dams, which have boosted economic development, have mostly been built in developing countries. Most industrialized countries have already developed appropriate sites.

Building reservoirs raises environmental, economic, health, and social issues and concerns. Two important issues include the displacement of floodplain residents and the loss of the most fertile and useful land in a given area. The potentially serious social consequences of displacing populations that may live on the floodplain must also be considered, as well as the environmental and economic costs of losing the land for hydropower purposes. In some areas, threats to endangered species—both animals and plants—may occur and need to be dealt with as well.

ENERGY FROM BIOMASS

Another source of indirect solar energy comes from plant biomass (such as woody, nonwoody, processed waste, or processed fuel) or animal biomass. Plants use solar energy during *photosynthesis* and store it as organic material as they grow. Burning or gasifying the resulting biomass reverses the process and releases the energy, which can then be used to generate heat or electricity or provide fuel for transportation.

Biomass has been used throughout history—burning wood in a campfire is burning plant biomass. Ancient cultures have used it for thousands of years for cooking and heating. Today, the global average is 10 to 14 percent of energy use is from biomass. It is higher in developing countries, however, ranging from 33 to 35 percent up to 90 percent in the poorest of countries. In primitive areas, only 10 percent of the energy in wood is captured and turned into usable energy, making it very inefficient. In developed countries such as Scandinavia, Germany, and Austria, they have the technology to use domestic biomass–fired heating systems and are able to achieve efficiencies of up to 70 percent with strongly reduced atmospheric emissions.

Biomass is also used to generate electricity commercially in many areas of the world. Commonly referred to as biopower, there are four basic types of biopower systems:

- direct-fired
- co-fired
- gasification
- small, modular systems

Most of the biopower plants in the world use direct-fired systems. They burn biomass *feedstock* directly to produce steam, which is captured by a turbine and then converted into electricity by a generator. The steam can also be used in various manufacturing processes. In Thailand, Indonesia, and Malaysia, for example, wood scraps from lumber and paper industries are fed directly into boilers to produce steam for manufacturing processes and to heat buildings.

Gasification systems use high temperatures and an oxygen-starved environment to convert biomass (usually wet organic domestic waste, organic industrial wastes, manure, and sludge) into a gas comprised of a mixture of hydrogen, carbon monoxide, and methane. The gas then fuels a gas turbine, which turns an electric generator. For large-scale gasification projects, the gas is thoroughly cleaned prior to its combustion.

When biomass decays in landfills, it produces methane, which can also be burned in a boiler to produce steam for electricity generation or for industrial processes. Wells are drilled into the landfill in order to recover the methane. Once the methane is recovered, pipes carry the gas to a central point where it is filtered and cleaned before burning.

Small modular systems can be either direct-fired, cofired, or gasification systems that generate electricity at a capacity of five megawatts or less. These systems are usually ideal in small towns or individual households.

Biomass is the only renewable energy source that can be converted directly into liquid fuels—called biofuels—for transportation purposes. The biofuels produced most often are ethanol and biodiesel. Ethanol is an alcohol made by fermenting biomass high in carbohydrates. These include substances such as sugarcane, maize, and corn. Ethanol is used mainly as a fuel additive to cut down a vehicle's carbon monoxide and other smog-causing emissions. Currently, Brazil operates the world's largest commercial biomass use program.

Biodiesel is an ester, which is similar to vinegar. Vegetable oils, animal fats, algae, and recycled cooking greases are used to produce it. It is used primarily as a diesel additive to reduce vehicle emissions or in its pure form to fuel a vehicle directly. Other biofuels include methanol and reformulated gasoline components. Methanol is produced through the gasification of biomass. After gasification, a hot gas is sent through a tube and then converted into liquid methane. Most reformulated gasoline components produced from biomass are pollution-reducing fuel additives, such as methyl tertiary butyl ether (MTBE) and ethyl tertiary butyl ether (ETBE).

Biomass can also be chemically converted into liquid, gaseous, and solid fractions by a process called pyrolysis, which occurs when biomass is heated in the absence of oxygen. This produces pyrolysis oil, which can be burned like petroleum to generate electricity. Pyrolysis oil is easy to transport and store and can be refined just as petroleum oil can. A chemical called phenol can also be extracted from pyrolysis oil, which can be used to make other products, such as wood adhesives, molded plastic, and foam insulation. Currently, other industrial uses of biochemicals are being researched. The DOE is conducting research on how to convert waste from landfills into biodegradable products.

Although biomass only captures roughly 1 percent of the Sun's available energy, it is attractive as an energy source because it can be easily stored for future use. Current advances in technology are increasing the efficiency with which the stored energy in biomass is converted to useable forms. The downside of using biomass is that it creates competition for an already limited supply of agricultural land. Critics also believe it will increase demand on water and soil resources, use agrochemicals, and threaten biodiversity.

A partial solution to these problems is to grow and harvest biomass crops sustainably. For example, perennial grasses such as switchgrass or elephant grass can actually help control erosion. Instead of devoting entire fields to biomass stock, these crops can be grown in between other crops on existing fields, which can actually be beneficial to the ecosystem. Some experts at DOE see a significant role for biomass energy use in the future.

In the United States, 45 billion kilowatt-hours of electricity is already being produced from biomass, which equals about 1.2 percent of the nation's total electric sales. In addition, almost 4 billion gallons (15 billion l) of ethanol are being produced—about 2 percent of the liquid fuels used in cars and trucks. According to the UCS, the contribution for heat is also substantial, but with better conversion technology and more attention paid to energy crops, the nation could produce much more. The DOE believes that the United States could produce 4 percent of its transportation fuels from biomass by 2010 and as much as 20 percent by 2030. For electricity, they estimate that energy crops and

BIOFUEL CROP BANS IN EUROPE

The European Union (EU), a 27-nation bloc, may impose a ban on the importation of fuels derived from crops that are grown on certain types of land—such as forests, wetlands, or grasslands. The law would not only ban those imports, it would also require the biofuels have a minimum level of greenhouse gas savings.

The crop used for biofuel in Europe is canola (also called rapeseed). Europe also imports palm oil from Southeast Asia and ethanol from Brazil. The ban would most likely affect the palm oil and Latin American imports.

Several recent studies have discredited some of the claims made by biofuel producers that the fuels help reduce greenhouse gases by reducing fossil fuel use and growing CO_2-consuming plants. They claim that growing the crops and turning them into fuel can instead result in considerable environmental harm.

The problem in Southeast Asia comes from the process that originates the biofuels. The environment is harmed in order to obtain them. *Peat* land areas are drained and deforested in order to plant palm plantations, which according to Adrian Bebb of Friends of the Earth, presently account for up to 8 percent of global annual CO_2 emissions.

In other areas, where native vegetation is being removed in order to plant crops, fossil fuels such as diesel for tractors, are often used to farm the crops that are going to be used in the biofuels. In addition, the crops are grown using hefty amounts of nitrogen fertilizer, further adding to the problem of global warming and environmental harm because the fertilizers are made from natural gas and the crops consume large amounts of water.

According to Bebb, "The active draining and deforesting of peat lands in Southeast Asia in order to cultivate palm plantations accounts for about 8 percent of global annual CO_2 emissions. In Indonesia, more than 44 million acres (18 million ha) of forest have already been cleared for palm oil development. The developments are also endangering wildlife like the orangutan and the Sumatran tiger, and putting pressure on indigenous peoples who depend on the forests."

The Royal Society, a national science academy in Britain, also stated that there is a need to distinguish between types of biofuels and that there should be specific goals for emission reductions. John Pickett, head

(continues)

(continued)

of biological chemistry at Rothamsted Research in Britain, said, "Indiscriminately increasing the amount of biofuels we are using may not automatically lead to the best reductions in emissions. The greenhouse gas savings of each depends on how crops are grown and converted and how the fuel is used."

Scientists at the Smithsonian Tropical Research Institute have also warned that biofuel production can result in environmental destruction, pollution, and damage to human health. William Laurance, a staff scientist at the Institute, said, "Different biofuels vary enormously in how eco-friendly they are. We need to be smart and promote the right biofuels."

Experts do agree that certain types of fuels made from agricultural wastes hold great potential to effectively combat global warming and still supply an adequate energy source. It is imperative, however, that governments set and enforce standards for how the fuels are produced. Experts also agree that with its new proposal, Europe appears to be moving ahead of the rest of the world in the discriminating production of clean biofuels.

crop residues alone could supply as much as 14 percent of the nation's power needs.

In addition to environmental benefits, biomass offers many economic and energy security benefits. By growing fuels at home, the nation reduces the need to import oil and reduces its exposure to disruptions in that supply. Farmers and rural areas gain a valuable new outlet for their products. Biomass already supports 66,000 jobs in the United States; if the DOE's goal is realized, the industry would support three times as many jobs.

OCEAN ENERGY

Oceans cover more than 70 percent of the Earth's surface. There are three basic ways to tap the ocean for its energy—high and low tides, wave action, and temperature differences. As the world's largest solar collectors, oceans generate thermal energy from the Sun. They also produce mechanical energy from the tides and waves. Even though the Sun affects all ocean activity, the gravitational pull of the Moon primarily drives the tides. And the wind powers the ocean waves.

Scientists and inventors have watched ocean waves explode against coastal shores, felt the pull of ocean tides, and desired to harness their incredible forces. As early as the 11th century, millers in Britain figured out how to use tidal power to grind their grain into flour. But it has only been in the last century that scientists and engineers have begun to look at capturing ocean energy to generate electricity.

Because ocean energy is abundant and nonpolluting, today's researchers are exploring ways to make ocean energy economically competitive with fossil fuels and nuclear energy. EU officials estimate that by 2010 ocean energy sources will generate more than 950 MW of electricity—enough to power almost 1 million homes in the industrialized world. Caused by the gravitational pull of the Moon and Sun and the rotation of the Earth, tides produce enormous, usable energy. Near shore, water levels can vary up to 40 feet (12 m). In order for tidal energy to work well, an area must be used that experiences a large diurnal change in tides. An increase of at least 16 feet (4.9 m) between low and high tide is needed. There are only a few places where this magnitude of tidal change occurs on Earth. Some power plants are already operating using this idea. For example, an ocean energy plant currently operating in France generates enough energy from tides to power 240,000 homes.

The simplest generation system for tidal plants involves a dam, known as a barrage, across an inlet. Sluice gates on the barrage allow the tidal basin to fill on the incoming high tides and to empty through the turbine system on the outgoing tide, also known as the ebb tide. There are two-way systems that generate electricity on both the incoming and outgoing tides. Tidal barrages can change the tidal level in the basin and increase turbidity in the water. They can also affect navigation and recreation. Potentially the largest disadvantage of tidal power is the effect a tidal station can have on plants and animals in the estuaries.

Tidal fences can also harness the energy of tides. A tidal fence has vertical-axis turbines mounted in a fence. All the water that passes through is forced through the turbines. They can be used in areas such as channels between two landmasses. Tidal fences have less impact on the environment than tidal barrages, although they can disrupt the movement of large marine animals. They are cheaper to install than tidal barrages. Tidal turbines are a new technology that can be used in many tidal areas. They are basically wind turbines that can be located anywhere there is strong tidal flow. Because water is about 800 times denser than air, tidal turbines have to be much sturdier than wind turbines. They will be heavier and more expensive to build but will be able to capture more energy.

Waves are caused by the wind blowing over the surface of the ocean. There is an incredible amount of energy in ocean waves. The total power of waves breaking around the world's coastlines is estimated at 2 to 3 million MW. The west coasts of the United States and Europe and the coasts of Japan and New Zealand are good sites for harnessing wave energy.

One way to harness wave energy is to bend or focus the waves into a narrow channel, increasing their power and size. The waves can then be channeled into a catch basin or used directly to spin turbines. Wave energy can be used to power a turbine. The rising water forces the air out of the chamber, and the moving air spins a turbine that can turn a generator. When the wave goes down, air flows through the turbine and back into the chamber through doors that are normally closed. Another type of wave energy system uses the vertical motion of the wave to power a piston that moves up and down inside a cylinder. The piston can also turn a generator, creating power. Most wave-energy systems today are small and can be used to power a warning buoy or a small lighthouse. Small, onshore sites have the best potential for the immediate future; they could produce enough energy to power local communities.

The energy from the Sun heats the surface water of the ocean. In tropical regions, the surface water can be 40°F (24°C) or more degrees warmer than the deep water. Using the temperature differences in ocean water to generate electricity is not a new idea. The idea dates back to the 1880s, when a French engineer named Jacques-Arsène d'Arsonval first developed the concept. Today, power plants can use the difference in ocean water temperatures to make energy. A difference of at least 38°F

(100.4°C) is needed between the warmer surface water and the colder deep ocean water to make this work.

One system—called the ocean thermal energy conversion (OTEC)—needs a temperature difference of at least 77°F (25°C) to operate, limiting its use to tropical regions. Hawaii has experimented with OTEC since the 1970s. There is no large-scale operation of OTEC today, because there are many challenges. First, the OTEC systems are not very energy efficient. Pumping the water is a serious engineering challenge itself. Electricity must also be transported to land. It will probably be 10 to 20 years before the technology is available to produce and transmit electricity economically from the OTEC systems.

Research is currently being done to place solar farms over the ocean. With oceans making up 70 percent of the Earth's surface, an ideal place for solar farms would be near the coasts. Currently, solar energy is used on offshore platforms and to operate remotely located equipment at sea. Along the coast of much of the United States, conditions are well suited to use wind energy. Currently, there is a plan to build a wind plant off the coast of Cape Cod, Massachusetts.

GLOBAL WARMING RESEARCH

One of the key areas being researched today is in carbon capture and sequestration. According to the DOE, before CO_2 can be sequestered from power plants and other point sources, it must be captured as a relatively pure gas. Existing capture technologies are not cost effective when trying to sequester CO_2 from power plants. For effective carbon sequestration, the CO_2 in the exhaust gases must first be separated and concentrated. Presently, CO_2 is captured from combustion exhaust by using cryogenic coolers. The current cost of CO_2 capture is around $150 per ton of carbon—a rate considered very costly. Based on an analysis conducted by SFA Pacific, Inc., it was shown that by adding the cost of existing technologies for CO_2 capture to an existing electricity generation process would increase the cost of electricity by 2.5–4 cents/kWh depending on the type of process.

Of the entire process—carbon capture, storage, transport, and sequestration—the capture portion represents about 75 percent of the total cost. Therefore, carbon capture research is being conducted and explored in an effort to reduce the cost. Sequestration methods are also being explored. The five options currently being researched by the DOE include:

- absorption (chemical and physical)
- adsorption (chemical and physical)
- low temperature distillation
- gas separation membranes
- mineralization and biomineralization

To date, several different options have been proposed that could reduce CO_2 capture costs. The DOE is currently looking at:

1. Research on improvements in CO_2 separation and capture technologies
   - New materials technologies—physical and chemical absorbents, carbon fiber, molecular sieves, polymeric membranes
   - Micro-channel processing units with rapid kinetics
   - CO_2 hydrate formation and separation processes
   - Oxygen enhanced combustion processes
2. Development of retrofittable CO_2 reduction—capture options for existing large point sources of CO_2 emissions such as electricity generation units, petroleum references, and cement and lime production facilities
3. Integration of CO_2 capture with advanced power cycles and technologies

Another area of research is the production of "green carbon." A small business in California is testing an alternative to carbon sequestration that takes waste CO_2 and tailings from mining operations and turns the material into a substance that can be used in a variety of industrial, agricultural, and environmental applications. The resulting substance is called precipitated calcium carbonate (PCC). PCC has been produced in the past via an energy-intensive process, but the green carbon technology transforms the carbon emissions instead of simply sequestering it. The PCC product can then be used in a variety of products, materials, and industrial processes. One of the biggest markets projected to use PCC is the paper industry as a filler and brightener. The industrial use of PCC is projected to grow to 10 million tons (9 million metric tons) by 2010.

The company spearheading all of this—Carbon Sciences—plans to take its research and implementation one step further: They plan to apply their green carbon technology at an ethanol plant, where the entire process will actually reduce the amount of CO_2, making the venture carbon negative (rather than even carbon neutral).

Another important area of global warming research concerns the role and contribution of non–CO_2 greenhouse gases. According to the Goddard Institute for Space Studies (GISS), some climate computer simulations about the future have led to the conclusion that Kyoto reductions will have little effect in the 21st century and that it may take "30 Kyotos" to reduce global warming to an acceptable level. Because of this, GISS has recommended research on, and the cutback of, non–CO_2 greenhouse gases and *black carbon* (soot) during the next 50 years. Based on GISS's research, non–CO_2 greenhouse gases have had the biggest impact on global warming. Cutting back on them, therefore, will help slow global warming.

The U.S. National Renewable Energy Laboratory (NREL) is currently involved in biomass research. Biochemical conversion technologies involve three basic steps: converting biomass to sugar or other fermentation feedstock, fermenting the product produced in the first step, and processing the fermentation product to yield fuel-grade ethanol and other fuels, chemicals, heat, and/or electricity. At NREL, researchers are trying to improve the efficiency and economics of the biochemical conversion process technologies by concentrating on simplifying the most difficult portions of the process. The current focus is on the pretreatment phase of breaking down hemicellulose to component cellulase enzymes, for breaking cellulose down to its component sugar. Researchers are also focusing on thermochemical conversion technologies that convert biomass to fuels, chemicals, and power using gasification and pyrolysis technologies. Gasification—heating biomass with about one-third of the oxygen necessary for complete combustion—produces a mixture of carbon monoxide and hydrogen, known

The State University of New York College of Environmental Science and Forestry (SUNY-ESF) biomass research farm in Tully, New York *(Lawrence P. Abrahamson, DOE/NREL)*

as syngas. Pyrolysis—heating biomass in the absence of oxygen—produces liquid pyrolysis oil. Both syngas and pyrolysis oil can be used as fuels that are cleaner and more efficient than solid biomass. Both can also be converted into other usable fuels and chemicals.

As research and discoveries continue and cleaner, more efficient energy sources are discovered and implemented, society advances closer to curbing global warming and its resultant harm to the environment.

Climate Modeling

The Earth's climate system is too complex for the human brain to grasp. There are so many interrelated forces constantly being influenced by outside factors and constantly shifting, trying to find some balance of equilibrium. It is simply not possible to write down a list of equations describing how the climate system works and reacts. The Earth's climate is not a straightforward process that gets from point A to point B every day in exactly the same way, at the same time, or in the same place. The only consistency about climate is that it is not consistent, and that is because there are so many variables involved and the patterns of possible interactions are enormous.

One of the key challenges climatologists face today with global warming is that it is important to be able to predict with some sense of confidence how the Earth's climate will change from region to region as temperatures rise so that policy makers can make appropriate decisions. Because of the inherent complexity and uncertainty, in order for climatologists to be able to do this they need to rely on climate models. Climate models are systems of differential equations derived from the basic laws of physics, fluid motion, and chemistry formulated to be solved on supercomputers.

This chapter discusses climate modeling—how it began, its fundamentals, and the challenges that both climatologists and computer programmers face today in its development. It also explores some of the diverse uses of climate models and how they are helping increase the scientific and public knowledge about global warming.

THE MODELING CHALLENGE—A BRIEF HISTORY

Climatology is a branch of physics, and physics makes use of two very powerful tools: experiments and mathematics. Weather and climate are so complex that without computers it would be impossible to mathematically quantify the climate system. Therefore, up until the computer age, there was no way to explain why and how climate behaved as it did. Once the technology developed, it was possible to build and assess quantitative climate models, because climate is based on physical principles.

The first objective of a climate model is to explain—however basically—the world's climates. Early on, the simplest and most widely accepted model of climate change was self-regulation, which means that changes are only temporary deviations from a natural equilibrium. Beginning in the 1950s, an American team began to model the atmosphere as an array of thousands of numbers. To answer the question about carbon, some primitive models were constructed representing the total carbon contained in an ocean layer, in the air, and in vegetation, with elementary equations for the fluxes of carbon between the reservoirs. Regardless of the carbon dioxide (CO_2) budget, scientists expected that natural *feedbacks* would operate and automatically readjust the system, restoring the equilibrium. Climatologists also recognized the need for more sophisticated models. They wanted to be able to explain triggers that caused past events, such as ice ages, plate tectonics, and changes in the ocean currents.

In the 1960s, computer modelers made encouraging progress by being able to make fairly accurate short-range predictions of regional weather. Modeling long-term climate change for the entire planet, however, was restricted because of insufficient computer power, ignorance of key processes such as cloud formation, inability to calculate the crucial ocean circulation, and insufficient data on the world's actual climate.

In the 1980s, models had improved enough that Syukuro Manabe, a senior meteorologist at Tokyo University, was able to use them to discover that the Earth's atmospheric temperature should rise a few degrees if the CO_2 level in the atmosphere doubled. Through the use of models, by the late 1990s, most experts acknowledged global warming and its effects. One area that scientists were interested in being able to model was that of climate surprises—rapid climate changes.

One of the most well-known models was an energy budget model developed by William Sellers of the University of Arizona in 1969. He computed possible variations from the average state of the atmosphere separately for each latitude zone. Sellers was able to reproduce the present climate and was able to document that it showed extreme sensitivity to small changes. He determined that if incoming energy from the Sun decreased by 2 percent (whether due to solar variation or increased dust in the atmosphere), it could trigger another ice age. Based on his results, Sellers suggested that "man's increasing industrial activities may eventually lead to a global climate much warmer than today."

Because an entire climate cannot be brought inside a laboratory, the only way to carry on an "experiment" of the entire system is to build a model of the entire system—a *proxy*. The most unpredictable part of the climate system—and as a result, one of the hardest to model—is the amount of radiation emitted by the Sun and the Earth. At any given time, water is present in water vapor, the oceans, and locked away in ice. The form and position the water takes change constantly in response to its interaction between solar and thermal radiation. Clouds (especially low-lying thick clouds) reflect huge amounts of sunlight back into space and keep it from overheating the Earth. High-altitude wispy clouds and water vapor absorb greater amounts of outgoing thermal (heat) radiation, which is generated off the Earth's surface after it gets warmed by the Sun.

In addition to greenhouse gases, clouds and water vapor contribute to keep the Earth's average temperature comfortably livable year round.

Atmospheric water has a tremendous effect on the Earth's climate. For years, researchers have been trying to understand all of the complex interactions: specifically, how clouds and water vapor will act if global warming escalates and the atmosphere gets hotter.

Scientists at the National Aeronautics and Space Administration (NASA) have currently developed several computer models to simulate the interactions between clouds and radiation. The area they are focusing most on is the Tropics because that region gets the most sunlight. Results so far have been mixed: Some say in the future low-lying thick clouds will increase, making global warming worse; others say when the Earth's surface heats up, cirrus clouds will dissipate and allow more thermal energy to escape to outer space.

The reason this is so difficult to model consistently is because clouds are constantly shifting, separating, growing, and shrinking. In addition, the only way to study them is through remote sensing (satellite imagery), which is still fairly new technology—satellites and image-processing software have only been around about 25 years.

Today, some of the "simple" models that can be run on desktop computers are comparable to what was once considered state of the art for even the most advanced computers in the 1960s. As a comparison, the computers used by NASA during the Apollo missions occupied an entire room. Today, those same programs can be run on a desktop computer. Computer models of the coupled atmosphere-land surface-ocean-sea ice system are essential scientific tools for understanding and predicting natural and human-caused changes in the Earth's climate.

FUNDAMENTALS OF CLIMATE MODELING

One of the key reasons climate is such a challenge to model is because it is a large-scale phenomena produced by complicated interactions between many small-scale physical systems. According to Gavin A. Schmidt at NASA's Goddard Institute for Space Studies (GISS), "Climate projections made with sophisticated computer codes have informed the world's policy makers about the potential dangers of anthropogenic interference with Earth's climate system. The task climate modelers have set for themselves is to take their knowledge of the local interactions of air masses, water, energy, and momentum, and from that knowledge explain the climate system's large-scale features, variability, and

Climate Modeling

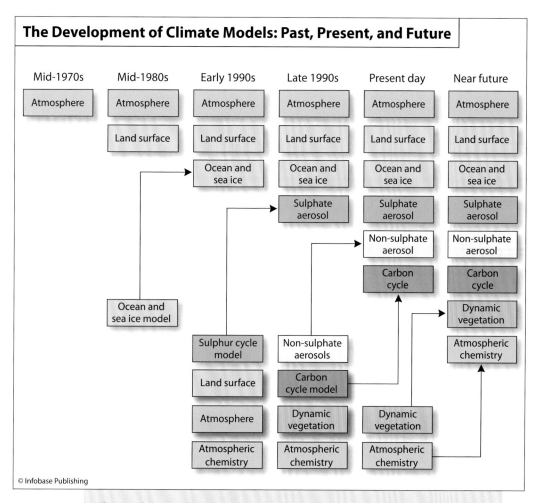

The evolution of climate models beginning in the mid-1970s and extending into the near future

response to external pressures, or 'forcings.' That is a formidable task, and though far from complete, the results so far have been surprisingly successful. Thus, climatologists have some confidence that theirs isn't a foolhardy endeavor."

It was not until the 1960s that electronic computers were able to meet the extensive numerical demands of even a simple weather system, such as low pressure and storm front. Since that time, more components have been added to climate models, making them more robust and complex, such as information characterizing land, oceans, sea ice, atmospheric aerosols, atmospheric chemistry, and the *carbon cycle.*

Models today are able to answer a wide range of questions, many geared specifically toward the effects of global warming.

The Physics of Modeling

The physics involved in climate models can be divided into three categories: fundamental principles (momentum, properties of mass, conservation of energy); physics theory and approximation (transfer of radiation through the atmosphere, equations of fluid motion); and empirically known physics (formulas for known relationships, such as evaporation being a function of wind speed and humidity).

Each model has its own unique details and will require several expert judgment calls. The most unique characteristic of climate models is that they have emergent qualities. In other words, when combining several interactions within the model, or parameters, the results of the interaction can produce an emergent quality unique to that system that was not previously obvious when looking at each system component by itself. For instance, there is no mathematical formula that describes the Earth's equatorial intertropical convergence zone (ITCZ) of tropical rainfall, which occurs through the interaction of two separate phenomena (the seasonal solar radiation cycle and the properties of *convection*). As more components are added to a model, it becomes more complex and can have more possible outcomes.

Simplifying the Climate System

All models must simplify complex climate systems. One critical aspect of climate models is the detail in which they can reconstruct the part of the world they are trying to portray. This level of detail is called *spatial resolution*. If a climate model has a spatial resolution of 155 miles (250 km), then there are data points draped around the globe like a net with an x/y/z coordinate set spaced on a grid at an interval of 155 miles (250 km). The z-coordinate—representing the vertical height—can vary, however. The resolution of a typical ocean model, for example, is 78–155 miles (125–250 km) in the horizontal (x/y) and 656–1,312 feet (200–400 m) in the vertical (z). Equations are generally solved every simulated "half hour" of a model run. Some of the smaller scale, localized processes such as ocean convection or cloud formation have to be generalized in a process called parametrization; otherwise it would be too demanding on the computer system.

There are three major types of processes that need to be dealt with when constructing a climate model: radiative, dynamic, and surface processes. Radiative processes deal with the transfer of radiation through the climate system, such as absorption and reflection of sunlight. In other words, where the sunlight travels once it is in the system. Dynamic processes deal with both the horizontal and vertical transfer of energy. This can include processes such as convection (the transfer of heat by vertical movements in the atmosphere, influenced by density differences caused by heating from below); diffusion (the spreading outward of energy throughout a system); and advection (the horizontal transport of energy through the atmosphere).

Surface processes are those processes that involve the interface between the land, ocean, and ice: the effects of *albedo* (how reflective a surface is); emissivity (the ability of a surface to emit radiant energy); and surface-atmosphere energy exchanges.

The simplest models have a "zero order" spatial dimension. The climate system is defined by a single global average. Models get more complex as they increase in dimensional complexity, from one-dimensional (1-D), to two-dimensional (2-D), to three-dimensional (3-D) models.

The complexity of the models is also controlled by changing the spatial resolution. In a 1-D model the number of latitude bands can be limited; in a 2-D model the number of grid points can be limited by spacing the points farther apart in a coarser grid. How long the model is run and the time intervals it is run on also affect the length and volume of the calculations involved.

Modeling the Climate Response

The purpose of a model is to identify the likely response of the climate system to a change in any of the parameters and processes, which control the state of the system. For example, if CO_2 is added into a simulation, the goal of the model is to see how the climate system will respond to it as the climate system tries to find an equilibrium. Or perhaps a model can focus on glacier melt and the results of ocean circulation as a result of the addition of freshwater and its effect on the climate.

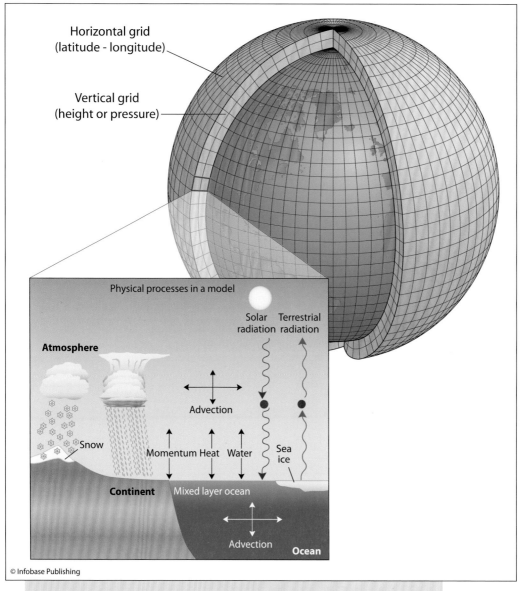

A climate model is comprised of a set of x/y/z points placed around the globe at specified intervals in a netlike structure, called its resolution. A small grid with lots of points close together has a high resolution and is more detailed; a large grid with points spread farther apart has a low resolution and less detail. In the model, each point x/y/z intersection has a value associated with it—one value for each variable represented in the model. In this example, each grid point would have a distinct value for solar radiation, terrestrial radiation, heat, water, advection, atmosphere, and so on.

Sometimes, complete processes can be omitted from a model if their contribution is negligible to the timescale being looked at. For instance, if a model is looking at a span of time that lasts only a few decades, there is no reason to model deep ocean circulation that can take thousands of years to complete a cycle. Not only would adding this data be useless, it would slow down the computer processing time and perhaps give erroneous results by trying to make a connection where none exists.

Types of Climate Models

There are several types of climate models, but they can be grouped into four main categories: *energy balance* models (EBMs); one-dimensional radiative-convective models (RCMs); two-dimensional statistical-dynamical models (SDMs); and three-dimensional general circulation models (GCMs). These four types increase in complexity from first to fourth, to the degree that they simulate particular processes, and in their temporal and spatial resolution. The simplest models do not allow for much interaction. The most complicated type—the GCM—allows for the most interaction. The type of model used depends on the purpose of the analysis. If a model is run that requires the study of the interaction between physical, chemical, and biological processes, then a more sophisticated model is normally used.

EBMs simulate the two most fundamental processes controlling the state of the climate—the global radiation balance and the latitudinal (equator to pole) energy transfer. Because EBMs are the most simplistic models, they are usually in a 0-D or 1-D format. In the 0-D form, the Earth is represented as a single point in space. In 1-D models, the dimension that is added is latitude; meaning that at whichever latitude interval is specified, the values in the model (such as albedo, energy flux, or temperature) would be input at each designated latitude.

RCMs can be 1-D or 2-D. Height is the attribute that is characteristic of these models. With the addition of the z-value, RCMs are able to simulate in detail the transfer of energy through the depth of the atmosphere. They can simulate the dynamic transformations that occur as energy is absorbed, scattered, and emitted. They can model and simulate the role and interaction of convection and how energy is transferred through vertical motion in the atmosphere. Also, because of their 2-D capability, they can simulate horizontally averaged energy transfers.

These models are helpful when climatologists are interested in understanding the fluxes between terrestrial and solar radiation that are constantly occurring throughout the atmosphere. When heat rates are calculated for different levels in the atmosphere, parameters such as cloud amount, albedo, and atmospheric turbidity are taken into account. The model can determine when the lapse rate exceeds its stability and convection (the vertical mixing of the air) takes place—a process called convective adjustment. RCMs are mainly used in studying forcing perturbations, which have their origin within the atmosphere, such as volcanic pollution.

SDMs are usually 2-D in form—a horizontal and vertical component. Currently there are many variations of them. These models usually combine the horizontal energy transfer modeled by EBMs with the radiative-convection approach of RCMs.

GCMs are sets of sophisticated computer programs that simulate the circulation patterns of the Earth's atmosphere and ocean. The models represent many complex processes concerning land, ocean, and atmospheric dynamics, using both empirical relationships and physical laws. By varying the amounts of greenhouse gases (GHGs) in the model's representation of the atmosphere, future climate can be projected both globally and regionally. GCMs cannot be used reliably, however, for scales smaller than a continent.

In the 1990s, GCMs began modeling the effects of aerosols in the atmosphere and scientists can now model GCMs for natural particulates (such as from volcanic eruptions) and anthropogenic aerosols from the burning of fossil fuels, sulfates, and organic aerosols through biomass burning. The purpose of GCMs is to describe how major changes in the Earth's atmosphere, such as changes in the GHG concentrations, affect climatic patterns including temperature, precipitation, cloud cover, sea ice, snow cover, winds, and atmospheric and ocean currents.

GCMs are not used to predict weather events, and their resolution is too coarse to predict the effects of local geographic features, such as specific mountains, that may influence climate. They have proven very useful, however, for examining long-term climatic trends, patterns, and responses to significant change. They are still notably complex when compared to the actual climate system though. According to the Met Office Hadley Centre, the foremost climate change research center in Britain, the table on page 170 illustrates the climate models they currently use.

Testing a Model—Modeling Trouble Spots

Models are tested at two different levels—at a small scale (did the wind patterns go in the right direction?), which includes the individual parameters; and at a large scale (did the atmosphere warm up?), where the predicted emergent features can be assessed.

The best way to test a climate model is to hindcast it—testing the model to see if it can forecast changes in climate that have already occurred. This is accomplished by plugging in previously measured parameters, such as ocean temperature and solar variability from past years, and running it in the virtual atmosphere of the climate model. The model is run forward through the past and into the present to predict changes in other atmospheric parameters—such as clouds and radiation balance. Ideally, the model should come up with the same values for clouds and radiation balance that are known to exist.

The 1991 eruption of Mount Pinatubo in the Philippines provided a good laboratory for model testing. Not only was subsequent global cooling of 0.8°F (0.5°C) accurately forecast soon after the eruption, but the radiative, water vapor, and dynamic feedbacks included in the models were quantitatively verified. This is as close to a controlled lab experience as global warming can get.

According to NASA, there are currently over a dozen facilities worldwide that are developing climate models. Over the past 20 years, the models have progressively become more sophisticated. Although errors overall between them appear to be unbiased, there are characteristics between the models that are similar, such as patterns of tropical precipitation.

Met Office Hadley Centre Model Configurations

	3-D atmosphere model (AGCM)	Atmospheric chemistry	Coupled atmosphere-ocean model (AOGCM) = AGCM + OGCM	Regional climate model (RCM)
ATMOSPHERE				
LAND SURFACE	AGCM plus "slab" ocean			
OCEAN	3-D ocean model (OGCM)	Carbon cycle		

Notes:

AGCM: *Atmosphere general circulation model. Consists of a three-dimensional representation of the atmosphere coupled to the land surface and cryosphere. AGCMs are useful for studying atmospheric processes, the variability of climate, and climate's response to changes in sea-surface temperature.*

AGCMs plus "slab" ocean: *This model predicts changes in sea-surface temperatures and sea ice by treating the ocean as though it were a layer of water of constant depth (usually 164 feet or 50 meters), heat transports within the ocean being specified and remaining constant while climate changes.*

OGCMs: *Ocean general circulation model is the ocean counterpart of an AGCM; a three-dimensional representation of the ocean and sea ice.*

Carbon cycle models: *The terrestrial carbon cycle is modeled within the land surface scheme of the AGCM, and the marine carbon cycle within the OGCM.*

Atmospheric chemistry models: *Three-dimensional global atmospheric chemistry models that look at the destruction of ozone and methane in the lower atmosphere.*

AOGCMs: *Coupled atmosphere-ocean general circulation models are the most complex models, consisting of an AGCM and an OGCM. Some models also include the biosphere, carbon cycle, and atmospheric chemistry.*

RCMs: *Regional climate models are those with resolutions of about 31 miles (50 km), designed to be used in smaller regional areas.*

WATCHING EARTH'S CLIMATE CHANGE IN THE CLASSROOM

NASA's GISS has developed an educational program that allows students to see how the Earth's climate is changing by being able to access NASA's global climate computer model (GCCM). It is giving students an opportunity to watch how a model takes data and calculates the amount of sunlight the Earth's atmosphere reflects and absorbs, the temperature flux of the atmosphere and oceans, the distribution of clouds, rainfall, and snow, and the dynamics of the world's ice caps.

While NASA scientists run the GCMs on supercomputers to simulate climate changes of the past and future, an educational version is being used by universities and high schools on desktop PCs. NASA's Educational Global Climate Model (EdGCM) was unveiled at the annual meeting of the American Meteorological Society in January 2005. The program is written so that students can conduct experiments similar to the ones scientists at NASA do.

According to Mark Chandler, lead researcher for the EdGCM project from Columbia University in New York City, "The real goal of EdGCM is to allow teachers and students to learn more about climate science by participating in the full scientific process, including experiment design, running model simulations, analyzing data, and reporting on results via the World Wide Web." In addition, an EdGCM cooperative is being designed to encourage communication between students at different schools and research institutions so that students can get a good idea of the role teamwork plays in scientific research today. The EdGCM also has a module devoted to global warming and CO_2 concentrations in the atmosphere, allowing students to analyze climate change. There is also a module on paleoclimate, enabling students to recreate climate conditions back when dinosaurs roamed the Earth.

Confidence and Validation

Although climate models should help clarify complex natural processes, the confidence placed in them should always be questioned. All climate models, by their very nature, represent a simplification of actual complicated processes. One thing that makes climate models so complex and difficult is that they often represent processes that occurred over timescales so long ago that it is impossible to test model results against real-world observations. Also, model performance can be tested through the simulations of shorter timescale processes, but short-term performance may not necessarily reflect long-term accuracy.

Because of the possibility of error, climate models must be used with caution, and the user must realize that a certain amount of uncertainty is present in the model. Margins of uncertainty must be attached to any model projection.

Validation of climate models (testing against real-world data) provides the only objective test of model performance. As an example, with prior GCMs, some validation exercises in the past have detected a number of deficiencies in various simulations, such as:

- Modeled stratospheric temperatures tended to be too low
- Modeled midlatitude westerlies tended to be too strong and easterlies too weak
- Modeled subpolar low-pressure systems in the winter tended to be too deep and displaced too far to the east
- Day-to-day variability tended to be lower than in the real world

Finding these discrepancies in models and correcting them are part of the process that enables the creation of stronger models. The process is iterative; no model is its strongest after the first run.

MODELING UNCERTAINTIES AND CHALLENGES

Because modeling is still in its infancy, its challenges are many. This section details the unknowns of modeling, including solar variability, the presence of aerosols, the characteristics of clouds, nature's unpredictability, error amplification, and other uncertainties.

Solar Variability

Solar variability is important in modeling climate. The total energy output of the Sun varies over time, causing warming and cooling cycles of the Earth's atmosphere. NASA satellites have confirmed that the Sun's energy output varies in sync with the 11-year sunspot cycle of magnetic changes in the Sun. Satellite data exist since the 1970s, giving climatologists only about 30 years of continuous data.

Climatologists can go farther back, however, and look at climate variations over centurylong intervals by analyzing the association of brightness changes with surface magnetic changes because records of the Sun's magnetism are available for several centuries back. Climatologists have records of lengths of sunspot cycles that are useful proxies as indicators of changes in the Sun's brightness. Comparisons can be calculated between sunspot cycle length and surface temperatures. Records have been constructed back to 1750.

The Sun's magnetic record can also be converted to estimate brightness changes and input into a climate simulation. According to scientists at the George C. Marshall Institute, using the Sun's magnetic records has shown that brightness changes have had a significant impact on climate change. Periods of a brighter Sun could contribute to warming of the Earth's atmosphere.

Aerosols

Pollutants such as sulfur dioxide make model predictions difficult. Aerosols form a haze that absorbs or reflects sunlight and causes a cooling effect, which offsets some of the predicted greenhouse warming. Aerosols can also change the properties and behavior of clouds. The theoretical effect of aerosols in modeling has been to cool the climate in both the present and the future. But so far, climatologists have had a difficult time getting models of aerosols to be consistent. Furthermore, as pollution issues are dealt with and aerosol content in the atmosphere diminishes, scientists need a solid understanding of their effect on global warming in order to be able to model changes associated with their reduction.

Clouds

Because clouds are a smaller-scale phenomena (they are generally smaller than the model's resolution) and transient—they come and go rather quickly—they are one of the most difficult properties to account for in climate models. One thing scientists are struggling with is how clouds will change in the future; specifically, how will their composition, structure, and extent change as the Earth's surface continues to get hotter.

Cloud behavior is extremely difficult to predict because there are so many variables that constantly change over time and space, such as surface temperature, air temperature, wind currents, varying amounts of water vapor, and abundance of aerosol particles.

According to NASA, all meteorological models inevitably fail at some point due to the sheer complexity of the Earth's system. To support this, *chaos theory* shows that weather will never be predictable with any significant accuracy for longer than two weeks, even with a nearly perfect model and nearly perfect input data. Today, climate models are still in their early stages of development—similar to the status of weather prediction 30 years ago.

Clouds have a very important role to play in climate models so climatologists are trying to understand their dynamic nature, enabling them to better accommodate them in models. According to NASA, clouds are the critical arbiters of the Earth's energy budget. Clouds cover 60 percent of the planet at any given time; they play a major role in how much sunlight reaches the Earth's surface, how much is reflected back into space, how and where warmth is spread around the globe, and how much heat escapes from the surface and atmosphere back into space. This makes clouds a key component of the Earth's climate system, and until scientists understand cloud physics better they will not be able to construct accurate global climate models.

Scientists at NASA have discovered that some clouds cool the surface by reflecting sunlight, and other types warm the surface by allowing sunlight to pass through and then trap the heat radiated by the surface. This proves there is a physical feedback loop between sea-surface temperature and cloud formation—each influences the other. Concerning global warming, a key question for climatologists and modelers is, "How will tropical clouds change if tropical sea-surface temperatures warm significantly?" One research team came up with a hypothesis that the Earth has a built-in mechanism for changing the structure and distribution of certain types of clouds in the Tropics to release more radiant energy into outer space as the surface warms.

One concept that has been proposed is called the Iris hypothesis. NASA uses remote-sensing satellites to obtain global measurements of the amount of sunlight reflected on the Earth and the amount of heat emitted up through the top of the atmosphere to calculate the bottom line on the Earth's energy budget. By doing this, scientists can determine which components of the Earth's system are most responsible for climate change. In the early 1980s, Richard Lindzen, a theoretician and professor of meteorology at the Massachusetts Institute of Technology (MIT), was interested in modeling how climate responds to changes in water vapor and cloud cover. He began looking closely at the presence of water vapor as a greenhouse gas and the effect it was having on global warming. The warmer the atmosphere becomes, the more water vapor it can hold. As the atmosphere absorbs CO_2 and the temperature rises, the additional heat allows the atmosphere to absorb even more water vapor. The water vapor further enhances the Earth's greenhouse effect in a progressive cycle. NASA scientists estimate that doubling the levels of CO_2 in the atmosphere are comparable to a 13 percent increase in water vapor. In the Tropics, clouds moisturize the air around them, and clouds are a major source of moisture.

Lindzen and his researchers focused on cloud cover using the Japanese Geostationary Meteorological Satellite-5 (GMS-5; Japanese name Himawari-5) to collect their measurements. The area they focused on was the area bordered by the Indonesian archipelago, the center of the Pacific Ocean, Japan, and Australia, because the area contains the world's largest and warmest body of water called the Indo-Pacific Warm Pool. What Lindzen wanted to determine was what type and extent of clouds are correlated to what ranges in sea-surface temperature. Lindzen said, "We wanted to see if the amount of cirrus associated with a given unit of cumulus varied systematically with changes in sea-surface temperature. The answer we found was, yes, the amount of cirrus associated with a given unit of cumulus goes down significantly with increases in sea-surface temperature in a cloudy region."

What they discovered was that the Earth has a natural adaptive infrared iris—a built-in check and balance mechanism that may be able to counteract global warming to some extent. Similar to the way the iris in a human eye contracts to allow less light to pass through the pupil under bright light, the iris hypothesis suggests that an area covered by high cirrus clouds contracts to allow more heat to escape into outer space when the environment gets too warm.

Although Lindzen is still trying to figure out exactly how the process works, his hypothesis is that the amount of cirrus precipitated out from cumulus depends upon what percent of the water vapor that is rising in a deep convective cloud condenses and falls as rain drops. Most of the water vapor condenses, but not all of it rains out. Some of the moisture rises in updrafts and forms thin, high cirrus clouds. Lindzen feels his discovery is important because if the amount of CO_2 is doubled in the atmosphere but there is no feedback within the system, then there is only 1 degree of warming. But climate models predict a much greater global warming because of the positive feedback of water vapor. What needs to be added to the model is the negative feedback (the infrared iris), which can be anywhere from a fraction of a degree to one degree—the same order of magnitude as the warming.

Not all scientists agree with Lindzen's model, and other scientists have not been able to reproduce it. It has garnered some attention, however. As more data are collected and more models are run, if repeatable results are obtained, then his theories may be pursued further.

Nature's Inherent Unruly Tendencies

According to Dr. Orrin H. Pilkey, a coastal geologist and emeritus professor at Duke, and Dr. Linda Pilkey-Jarvis, a geologist at Washington State Department of Geology, depending too much on computer models may not be completely reliable because "nature is too complex and depends on too many processes that are poorly understood or little monitored—whether the process is the feedback effects of cloud cover on global warming or the movement of grains of sand on a beach."

One thing they criticize about mathematical models is that there are too many fixed mathematical values applied to phenomena that change often. Another modeling weakness is that formulas may include coefficients (also called fudge factors according to Dr. Pilkey) to ensure that they come out right. In addition, sometimes modelers fail to verify that a project performed as predicted, considering nature's possible unruly outcomes. On the other hand, Dr. Pilkey also cautions against moving too far in the other direction, especially when modeling climate change. According to him, "Experts' justifiable caution about model uncertainties can encourage them to ignore accumulating evidence from the real world."

The Pilkeys also stress "It is important to remember that model sensitivity assesses the parameter's importance in the model, not necessarily in nature. If a model itself is a poor representation of reality, then determining the sensitivity of an individual parameter in the model is a meaningless pursuit."

What they suggest, perhaps alongside, if not in replacement of, is adaptive management. With this approach, policy makers can start with a model of how an ecosystem works but make constant observations in the field, altering their policies as conditions change. The problem with this approach is that because of management, funding, and policy issues, these requirements are often hard—if not impossible—to achieve. When models are used, they do have some basic recommendations for how to better use them: pay more attention to nature to accumulate information on how living things and their environments interact, modelers should state explicitly what assumptions they have made; modelers should seek to discern general trends instead of giving a model more analytical power than it probably has; and models should be complemented with observations from the field.

According to Dr. Pilkey, "If we wish to stay within the bounds of reality we must look to a more qualitative future—a future where there will be no certain answers to many of the important questions we have about the future of human interactions with the Earth."

Error Amplification

If a compass heading is set even a half degree off, the farther the boat travels, the farther off course it becomes, the error growing in magnitude the longer the boat progresses. In large, complex models, such as GCMs, if there is an initial input error—however tiny—in the physics of climate data, as the model runs, it can accumulate, adding up through the millions of numerical operations to give an impossible final result. This can render a model completely useless if the error is not initially caught and fixed.

One approach in fine-tuning large climate models is to construct simpler models of the interactions between biological systems and gases. By improving the interactions of the individual components within the system, potential errors can be culled out and corrected before being added to a large model where even a small measurement can eventually become amplified into a major error.

Modeling Uncertainties and Drawbacks

One of the biggest drawbacks climate modelers face today is that direct, observational data is extremely limited. Global temperatures have only been collected and monitored for about 100 years. Many climate modelers believe climate modeling is still in its infancy and with many hurdles to overcome, not only in the mathematics of modeling itself and computer development, but also in understanding climate processes themselves. In some areas, uncertainties have actually grown.

Some of that uncertainty is reflected in the comments of three climate modelers: Gerald North of Texas A&M University says, "The uncertainties are large." Peter Stone of MIT says, "The major climate prediction uncertainties have not been reduced at all." The cloud physicist Robert Charlson, professor emeritus at the University of Washington, Seattle, says, "To make it sound like we understand climate is not right."

Stone takes it further when referring to the "politically charged atmosphere" of global warming today and the fact that the inherent uncertainties in modeling are being focused on and used as fuel to dismiss them because possibly they are making global warming appear worse than it is. He comments, "We can't fully evaluate the risks we face. A lot of people won't want to do anything. I think that's unfortunate. Greenhouse warming is a threat that should be taken seriously. Possible harm could be addressed with flexible steps that evolve as knowledge evolves. By all accounts, knowledge will be evolving for decades to come."

Climate modeling has three basic challenges to improve accuracy: detecting consistently rising temperatures, attributing that warming to rising greenhouse gases, and projecting warming into the future.

Michael Mann, a climatologist at the University of Virginia, said the first challenge has already been resolved by the Intergovernmental

Panel on Climate Change (IPCC) in their 2007 report. He credits part of their increased confidence to more sophisticated and effective statistical techniques for analyzing sparse observations.

Concerning the rising GHG challenge, David Gutzler of the University of New Mexico says, "Attributing the warming to greenhouse gases is much harder. To pin the warming on increasing levels of greenhouse gases requires distinguishing greenhouse warming from the natural ups and downs of global temperature."

The IPCC's 1995 report said data "suggested" a human influence toward the rising GHGs. In their recent report, however, their attribution statement was much stronger: ". . . most of the observed warming over the last 50 years is likely (66–90 percent) to have been due to the increase in greenhouse gas concentrations."

The climate modeler Jerry D. Mahlman, the recently retired director of the National Oceanic and Atmospheric Administration's (NOAA) Geophysical Fluid Dynamics Laboratory in Princeton, New Jersey, comments on the IPCC's 2007 report, "I'm quite comfortable with the confidence being expressed. The report states that confidence in the models has increased. Some of the model climate processes, such as ocean heat transport, are more realistic; some of the models no longer have the fudge factors that artificially steadied background climate; and some aspects of model simulations, such as El Niño, are more realistically rendered. The improved models are also being driven by more realistic climate forces. A Sun subtly varying in brightness and volcanoes spewing sun-shielding debris into the stratosphere are now included whenever models simulate the climate of the past century."

According to Mahlman, other modeling uncertainties that still need to be improved include the role of atmospheric aerosols, lack of enough data, cloud behavior, anthropogenic effects, global cooling, future pollution control, and future social behavior.

Jeffrey Kiehl of the National Center for Atmospheric Research (NCAR) in Boulder, Colorado, says, "A number of uncertainties are still with us, but no matter what model you look at, all are producing significant warming beyond anything we've seen for 1,000 years. It's a projection that needs to be taken seriously."

Other Unknowns

Other current modeling challenges include the carbon cycle, future economics, past and future temperatures, cooling effects, abrupt weather events, and future thermohaline circulation. The direct effect of CO_2 on global warming is presently accounted for in current models, but what needs better clarification is to what extent CO_2 influences global temperatures due to its secondary influences. For example, models still need to determine how much of the anthropogenic CO_2 actually makes it into the atmosphere. Scientists know that not all human-attributed CO_2 emissions end up in the atmosphere; some are absorbed by the Earth's natural carbon cycle and end up in the oceans and terrestrial biosphere (plants, soils) instead. Because the Earth's carbon cycle is extremely complicated, scientists still need to better understand how the carbon sources and sinks work in the cycle in order to enable climate models to better represent that attribute.

Another problem is trying to predict future CO_2 emissions since they will be influenced by worldwide growth patterns. The role of developing countries and their fossil fuel use will become critical, as will the rate at which countries switch to renewable energy sources. The enforcement of pollution controls and the rate of deforestation will have effects that are difficult to predict.

Temperature is also a difficult variable to determine. Future global temperature is difficult to predict because the atmosphere is so sensitive to the concentration of aerosols and CO_2. Because of this sensitivity, even small input errors can accumulate into misleading modeling results. The cooling effects from particles in the atmosphere, such as aerosols, sulphur emissions, and volcanic eruptions, can have significant local or regional impacts on temperature. In models this can affect albedo and reflection values. To help manage for this, global cooling parameters may need to be added to the model. Abrupt weather events are not currently predictable because present-day models' spatial resolutions are too coarse. As an example, in some climate models, New Zealand is only represented by 10 data points—not nearly enough resolution to study small-scale spatial events like changes in air currents.

Modeling of the thermohaline circulation (ocean conveyor belt) faces uncertainties due to the complexities controlling deepwater for-

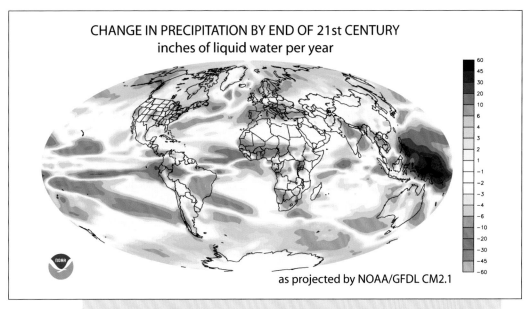

Based on a model produced by NOAA, this graphic illustrates one model's prediction of future global precipitation trends by the end of the 21st century. *(NOAA)*

mation, the interrelationship between large-scale atmospheric forcing with warming and evaporation at low latitudes, and cooling and increased precipitation at high latitudes. Uncertainty also lies in trying to model the addition of freshwater from the Arctic to the tropical Atlantic. Rates and direction of flow and convection are extremely difficult to predict at this point. According to NASA, other challenges are extreme events such as hurricanes and heat waves, the turbulent behavior of the near-surface atmosphere, and the effects of ocean eddies. Concerning climate models overall, the NASA scientist Gavin Schmidt says, "Climate models are unmatched in their ability to quantify otherwise qualitative hypotheses and generate new ideas that can be tested against observations. The models are far from perfect, but they have successfully captured fundamental aspects of air, ocean, and sea-ice circulations and their variability. They are, therefore, useful tools for estimating the consequences of humankind's ongoing and audacious planetary experiment."

Practical Solutions That Work—Getting Everyone Involved

Increasing human consumption of natural resources is at the root of several of the global environmental problems faced today. As the rate of consumption of natural resources increases, these resources become stressed, contributing to global warming and the wastes and pollution that are created as a result. This threatens both the health and quality of life of people and ecosystems worldwide. The unsustainable consumption and waste production patterns, whether water use, GHG emissions, or other activities, have effects that reach the entire planet. Environmental and human health are affected globally. Every person's ecological footprint changes the environment, and the exact size of that footprint is determined by an individual's actions and choices concerning recycling, fossil fuel consumption, food choices, or other lifestyle choices that can hurt the global ecosystem. This chapter discusses a multitude of different, simple ways that communities and individuals can get involved in fighting global warming.

TAKING ACTION

Action can be taken at many levels. Oftentimes people do not think taking action on an individual level will make a big enough difference. But when many individuals make similar decisions, then collectively huge differences can be realized. One example of this can be seen in recycling programs. Twenty years ago, hardly anyone recycled in the United States; people threw items in the garbage without a second thought. Slowly, an environmental movement took hold, however, and as people became educated about the positive benefits of recycling, most households today have gone from recycling nothing to recycling newspapers, aluminum, plastics, glass, and other metals. Ten years ago, if someone was asked what the color green meant to them, they might have said jealousy. Today, they would more likely say environmentally friendly. In addition, many businesses participate in recycling programs, buy recycled products, and practice source reduction in their packaging efforts. People today have a much different way of thinking than they did a generation ago. The same phenomenon can take place with the issue of global warming.

Taking action on global warming may only take a small change in lifestyle and behavior to make a big change in GHG reductions. When that action is multiplied by 6.8 billion people worldwide, the changes can be significant. Businesses and communities can make a difference too. As consumers, people have many commercial choices and can support businesses that promote a sustainable future, rather than those that do not take care of the environment. Consumers need to support businesses that use clean, efficient technology. Corporations also have a responsibility to conserve natural resources and protect community health, which should be reflected in the decisions they make. Everyone can encourage community action. Simple things, such as urging local businesses and community buildings to install bike racks; promoting community car-pooling plans and the construction of bike lanes; working to change local zoning ordinances and other regulations that involve energy use; and encouraging local electric utilities to promote energy efficiency and the use of clean, renewable energy sources, can make a significant difference.

It is also possible to take national action by writing to local newspapers about the significance of the global warming threat and the need for U.S. leadership. Contacting congressional representatives is another way to be heard and encourage Washington to take proper action against global warming. Asking governors, legislators, and utilities to promote energy efficiency, nonpolluting transportation alternatives, and the development of clean renewable sources of energy is also vital. City planners should be encouraged to ensure that buildings are built or remodeled with energy efficiency in mind.

Many companies today are taking the initiative to green up instead of waiting for government to mandate regulations. A *New York Times* article of January 21, 2008, stated that 11 companies had teamed up to see how they could coordinate with thousands of their suppliers to cut back GHG emissions. The companies participating in the program included some of the biggest corporations and was coordinated by a British organization called the Carbon Disclosure Project, which helps companies cooperate in the war against global warming. The goal of these companies was to find where in the supply chain GHGs were being emitted and what risks there were. In phase one, the major companies worked with 50 suppliers and then completed an evaluation and recommended where emissions could be cut. Phase two, in progress now, will include up to 2,000 suppliers associated with each company and GHG cutbacks will be designed. When businesses step up and become involved in curbing global warming, they set a good example and educate others.

PRACTICAL SOLUTIONS TO GLOBAL WARMING

Solutions to climate change are now available, and the Union of Concerned Scientists (UCS) thinks that they are practical and doable right now. The steps make sense and will even save consumers money. The cost of inaction, they warn, is unacceptably high. According to the UCS, procrastination is not an option—if aggressive action waits another 10 or 20 years global warming will have escalated to the point where it will not only be more difficult to deal with, but the consequences will be much more severe.

UCS likens society's current treatment of the atmosphere to the manner in which rivers were treated at one time. People used to dump waste into the waterways, believing the rivers had a place to handle it, never stopping to think about who might live downstream. Then, when entire

THE 2007 NOBEL PEACE PRIZE

The Norwegian Nobel Committee decided that the Nobel Peace Prize for 2007 was to be shared equally between the Intergovernmental Panel on Climate Change (IPCC) and Albert Gore, Jr., for their efforts to raise awareness and spread knowledge about man-made climate change and to lay the foundation for the measures that are needed to counteract such change.

The Nobel Committee acknowledges that climate change may alter and threaten the living conditions of mankind; it may trigger large-scale migration and lead to competition for the world's limited resources. Such serious changes will place extreme burdens on the most vulnerable countries, possibly triggering violence and wars. Over the past two decades, the IPCC has created scientific reports about the connection between human activities and global warming. Thousands of scientists have worked long hours to achieve greater certainty as to the magnitude of the warming. The IPCC has made significant progress in furthering an understanding of climate change as well as finding additional evidence to support its existence.

Al Gore is credited with having long been regarded as one of the world's leading environmentalist politicians. His strong commitment has been demonstrated in his political activity, films, books, and lectures. His contribution has greatly strengthened the struggle against climate change. The committee considers him the single individual who has done most to create greater worldwide understanding of the measures that need to be adopted.

By awarding the Nobel Peace Prize for 2007 to the IPCC and Al Gore, the Norwegian Nobel Committee sought to contribute to a sharper focus on the processes and decisions that are necessary to protect the world's future, and thereby reduce the threat to the security of mankind. Commenting on the award, Gore said, "Climate change is a real, rising, imminent, and universal threat to the future of the Earth. Our world is spinning out of kilter. We, the human species, are confronting a planetary emergency—a threat to the survival of our civilization that is gathering ominous and destructive potential."

In a speech Gore said, "The future is knocking at our door right now. Make no mistake—the next generation will ask us one of two questions. Either they will ask, "What were you thinking: why didn't you act?" Or they will ask instead, "How did you find the moral courage to rise and successfully resolve a crisis that so many said was impossible to solve?"

Source: Norwegian Nobel Committee

fisheries were poisoned and some rivers even caught on fire because they were so polluted, society finally paid attention to the heavy environmental impact and changed the laws. Today, this is the stage where the Earth's atmosphere is. Every ton of CO_2 added each day to the atmosphere will stay there for about 100 years—directly affecting the next three to four generations. As CO_2 continues to be added, the time line keeps extending farther and farther. The UCS has identified the five following commonsense climate change solutions that could be put into effect today.

Commonsense Solution #1: Make Better Cars and SUVs

The technology currently exists to build cars, minivans, and SUVs that get 40 MPG and more. The technology for better transmissions and engines exists, aerodynamic designs can be easily altered, and stronger yet lighter material can increase the average fuel economy of today's automotive fleet from 24 MPG to 40 MPG over the next 10 years. If these changes were made, it would be the equivalent of taking 44 million cars off the highways and it would save drivers thousands of dollars in fuel costs. According to the UCS, the transportation sector accounts for almost 30 percent of U.S. annual CO_2 emissions. Therefore, raising fuel economy is one of the most significant areas to focus on changing. Positive steps can be taken in several areas: Manufacturers need to offer cars with better mileage, the federal government can offer new-technology vehicles (such as hybrids), and consumers can purchase cars with the best fuel mileage.

A step in the right direction was the Car Allowance Rebate System (CARS) program—also called the Cash for Clunkers program—that was put into effect during summer 2009. Signed into law by President Obama on June 24, 2009, it remained in effect until August 24, 2009. The goal of the program was to encourage consumers to trade in older, less fuel-efficient vehicles for new vehicles that get better fuel economy by providing a credit of either $3,500 or $4,500. The program was modeled after others that have been successfully run in Europe.

The program divided cars, trucks, SUVs, and vans into four categories, usually based on weight and length of their wheelbase. The vehicles that were traded in will be destroyed (not resold), and the base manufacturer's suggested retail price of the new replacement vehicle could not exceed $45,000 in order to qualify. The MPG figures used in the trade-ins were taken from the EPA's published "combined" MPG tables.

In order to be eligible, the trade-in car had to be in drivable condition, registered and insured consistent with state law, be less than 25 years old, and have a combined MPG of 18 or less. The car being acquired had to be a new model with a base manufacturer's suggested retail price of $45,000 or less. When the program officially came to a close, nearly 700,000 clunkers had been taken off the highways, replaced by far more fuel-efficient vehicles. Rebate applications worth $2.877 billion had been submitted by the deadline, under the $3 billion provided by Congress to run the program. Initially, the program was supposed to run until November 2009, but the program was so successful, the designated funds were depleted much faster. Cars manufactured in the United States topped the most-purchased list, including the Ford Focus, the Honda Civic, and the Toyota Corolla.

According to U.S. transportation secretary Ray LaHood, "American consumers and workers were the clear winners thanks to the Cash for Clunkers Program. Manufacturing plants have added shifts and recalled workers. Moribund showrooms were brought back to life and consumers bought fuel-efficient cars that will save them money and improve the environment. This is one of the best economic news stories we've seen and I'm proud we were able to give consumers a helping hand."

According to DOT news bulletin 133-09, the program also benefited the economy as a whole. Based on calculations by the White House Council of Economic Advisers, the CARS program will boost economic growth in the third quarter of 2009 by 0.3–0.4 percentage points at an annual rate thanks to increased auto sales in July and August. It will also sustain the increase in gross domestic product (GDP) in the fourth quarter because of increased auto production to replace depleted inventories. It will also create or save 42,000 jobs in the second half of 2009. Those jobs are expected to remain well after the program's close.

Both Ford and General Motors have announced production increases as a spin-off of the program. It also means good news for the environment: 84 percent of the consumers traded in trucks and 59 percent purchased passenger cars. The average fuel economy of the vehicles traded in was 15.8 MPG and the average fuel economy of the vehicles purchased was 24.9 MPG—a 58 percent improvement.

"This is a win for the economy, a win for the environment, and a win for American consumers," Secretary LaHood remarked.

The following tables illustrate some of the statistics reflected by the program.

Car Allowance Rebate System (CARS)	
Dealer Transactions: Number submitted: 690,114 Dollar value: $2,877.9 million	
Top 10 New Vehicles Purchased	
1	Toyota Corolla
2	Honda Civic
3	Toyota Camry
4	Ford Focus FWD
5	Hyundai Elantra
6	Nissan Versa
7	Toyota Prius
8	Honda Accord
9	Honda Fit
10	Ford Escape FWD
Top 10 Trade-in Vehicles	
1	Ford Explorer 4WD
2	Ford F150 Pickup 2WD
3	Jeep Grand Cherokee 4WD
4	Ford Explorer 2WD
5	Dodge Caravan/Grand Caravan 2WD
6	Jeep Cherokee 4WD
7	Chevrolet Blazer 4WD
8	Chevrolet C1500 Pickup 2WD
9	Ford F150 Pickup 4WD
10	Ford Windstar FWD Van

Vehicles Purchased by Category	
passenger cars	404,046
category 1 truck*	231,651
category 2 truck**	46,836
category 3 truck***	2,408

Note: *category 1 truck: SUVs, small and medium pickup trucks, minivans, and small and medium passenger and cargo vans
**category 2 truck: large pick-up trucks
***category 3 truck: very large vans, SUVs, pickups, and work trucks

Vehicle Trade-in by Category	
passenger cars	109,380
category 1 truck	450,778
category 2 truck	116,909
category 3 truck	8,134

Average Fuel Economy

New vehicles mileage: 24.9 MPG
Trade-in mileage: 15.8 MPG
Overall increase: 9.2 MPG, or a 58 percent improvement

Another exciting new development centers around the recent announcement by General Motors of their new Chevy Volt—to be released in 2011. The Volt is GM's all-electric car and is promising to get a staggering 238 MPG. According to Frank Weber, vehicle chief engineer for the Volt, the mileage rating is based on combined electric-only driving and charge-sustaining mode with the engine running. The car runs entirely on electric power stored up in its battery and has a range of 40 miles before a small gasoline engine starts adding additional electricity to the battery pack. As science and technology continue to experiment with hybrids, electric cars, fuel cells, and other technology, the benefits and discoveries will further benefit the consumer as well as the environment.

Commonsense Solution #2: Modernize America's Electricity System

Currently, more than half of America's electricity is produced from outdated, coal-burning power plants that dump pollutants and heat-trapping gases into the atmosphere. Cost-effective, clean energy sources do exist. The use of clean, renewable energy needs to be increased. If more investments were made in energy efficiency and in reducing pollution from fossil fuel plants, several direct benefits would be realized: Consumers would save money, heat-trapping emissions would be reduced, and the dependence on fossil fuels would be lessened or eliminated.

A study conducted by UCS stated that the United States could reduce power plant CO_2 emissions by 60 percent compared with government forecasts for 2020. Consumers would save a total of $440 billion—reaching $350 annually per family by 2020. UCS believes that a national standard requiring 10 percent of electricity in the United States to be generated from renewable energy is within reason. Areas around the country are already using wind, solar, geothermal, and biomass to produce energy. Costs have dropped significantly, as well. As an example, a kilowatt-hour of wind energy in 1980 was 40 cents. Today, it ranges from three to six cents.

The UCS suggests the establishment of a renewable electricity standard that requires utilities to generate 10 percent of their power from clean, renewable energy sources. UCS and the Energy Information Administration (EIA) analyzed the effects of a 10 percent mandatory use of renewable energy and determined CO_2 emissions would be reduced 183–237 million tons (166–215 million metric tons) nationally by 2020—the equivalent to taking 32 million cars off the road. This approach would also help the U.S. economy because the fuels would be produced in the United States, creating more than 190,000 jobs and providing $41.5 billion in new capital investment. To date, 20 states have already adopted standards requiring utilities to offer more renewable energy to customers.

Commonsense Solution #3: Increase Energy Efficiency

Technology is already available to create more efficient appliances, windows, homes, and manufacturing processes. These solutions are currently saving homeowners money and have a significant impact on the Earth's climate. The UCS has calculated that energy-efficient appliances have kept 53 million tons of heat-trapping gases out of the atmosphere each year. New or updated standards for many major appliances, including washers, dishwashers, water heaters, furnaces, boilers, and air-conditioners have been put in place to increase efficiency. By 2020, these efficiency gains alone will reduce the need for up to 150 new medium-sized (300 megawatt) power plants.

When replacing appliances, consumers should always look for the ones with the Energy Star label on them. If each household in the United States replaced its existing appliances with the most efficient models available, it would save $15 billion annually in energy costs and eliminate 175 million tons of heat-trapping gases.

Many utility companies offer free home energy audits. It often pays to take advantage of this service to discover ways to cut back on energy use. Simple measures, such as installing a programmable thermostat to replace an old dial-type unit or sealing and insulating heating and cooling ducts, can reduce a typical family's CO_2 emissions by about 5 percent.

Commonsense Solution #4: Protect Threatened Forests

In addition to providing a home for more than half of the Earth's species and providing benefits such as clean drinking water, forests also play a significant part in climate change. They store immense amounts of carbon. Unfortunately, when forests are burned, cleared, or degraded, the carbon that is stored in their leaves, trunks, branches, and roots is released into the atmosphere. In fact, tropical deforestation now accounts for about 20 percent of all human-caused CO_2 emissions each year.

In order to combat the effects of global warming, forested areas should be managed appropriately. In the United States, for example, the forests of the Pacific Northwest and Southeast could double their storage of carbon if timber managers lengthened the time between harvests and allowed older trees to remain standing. Conservation practices and incentives should also be extended to private companies. It would be helpful if a system was set up that allowed private companies to get credit for reducing carbon when they acquire and permanently set aside natural forests for conservation instead of using the land for another economic venture.

The UCS also recommends not clearing out mature forests to replace them with fast-growing younger trees in a tree plantation venture. Although younger trees do draw carbon out of the atmosphere more quickly, cutting down mature forests releases large quantities of CO_2 into the atmosphere. In addition, replacing natural forests with tree plantations destroys biodiversity.

Commonsense Solution #5: Support American Ingenuity

With prior achievements such as the Apollo program, the silicon chip, and the Internet, America has proven that putting together the best minds and the right resources can result in technological breakthroughs that change the course of human history. To date, federal research funding has played an integral part in the progress of developing renewable energy sources and improving energy efficiency. Over the past 20 years, the Department of Energy's efficiency initiatives have saved the country 5.5 quadrillion BTUs of energy and nearly $30 billion in avoided energy costs. Federal research dollars have driven technological advances in fuel cells. This technology, which runs engines on hydrogen fuel and emits only water vapor, is key to moving our transportation system away from the polluting combustion engine and freeing the United States from its oil dependence.

It will take continued and dedicated support for research and development to achieve the practical solutions needed to overcome global warming. According to UCS, far more is currently invested in subsidies for the fossil fuel and nuclear industries than on Research and Development for renewable energy or advanced vehicle technologies. For example, Congress appropriated $736 million for fossil fuel research and $667 million for nuclear research in 2001, but only $376 million for all renewable energy technologies combined. The President's Council of Advisors on Science and Technology recommended that double be spent on energy efficiency and renewable energy technologies. Vehicle research should also be increased and refocused on technologies and fuels that can deliver the greatest environmental gains, including hybrid and fuel cell cars, renewable ethanol fuel, and the cleanest forms of hydrogen production.

Another area where research money needs to be directed is in geologic carbon sequestration as a potentially viable way to reduce CO_2 released into the atmosphere. Even though this technology holds promise, it is still under development and its environmental impacts must be fully explored before it will be able to be widely used. The UCS believes the United States has a clear moral responsibility to lead the way internationally and has the financial and technical expertise that will help reap the economic benefits of new markets for clean technology exports.

SUGGESTED SOLUTIONS THAT ARE NOT SO PRACTICAL

In an effort to find solutions to stop or slow global warming, there have even been suggestions referred to as geoengineering solutions that seem a little futuristic and far-fetched. Five of the most often cited follow.

Copying a Volcano

One suggested solution is to copy a natural volcano. A violent volcanic eruption, such as that of Mount St. Helens in 1980, can eject millions of tons of sulfur dioxide gas into the atmosphere, creating a continuous cloud that blocks the Sun's radiation. Based on this principle, it has been suggested that by injecting the atmosphere with sulfur, it may be possible to block solar radiation and potentially cool the planet.

According to Alan Robock, an environmental scientist at Rutgers University, sulfur dioxide reacts with water in the atmosphere to create droplets of sulfuric acid, which function to scatter the Sun's light back out into space. One reason why sulfur dioxide has been suggested is because sulfur does not heat the stratosphere like other aerosols do, so in theory it would not work against the cooling effect. Another option would be hydrogen sulfide, but it would require an enormous amount in order to be effective. It would take five megatons each year to counteract the effects of global warming. Robock likens that to having the eruption of a volcano a quarter the size of Mt. Pinatubo every year.

Five Big Plans to Stop Global Warming

Copy a volcano

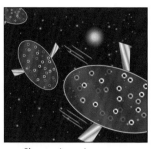
Shoot mirrors into space

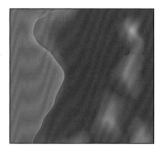
Seed the sea with iron

Whiten the clouds with wind-powered ships

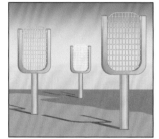
Build fake trees

Several geoengineering projects have been suggested to stop global warming, many sounding like something Jules Verne could create, including copy a volcano, shoot mirrors into space, seed the sea with iron, whiten the clouds with wind-powered ships, and build fake trees.

Robock cautions that there is no way to engineer a method to propel the sulfur upward into the atmosphere with the intensity and force of a volcano. Suggestions have been made that perhaps it can be launched by planes. The problem with that is that only small fighter jets can reach the stratosphere and they would not be able to carry enough particles of sulfur hydroxide to do it. Heavy artillery—shooting sulfur-laden cannonballs that would explode in the stratosphere—has also been suggested, as has sending balloons carrying gas, but so far nothing concrete has come out of it. Others argue that even if the balloon idea were technically feasible, there would be a problem when all the spent balloons fell back to Earth.

Seed the Sea with Iron

Another geoengineering scheme is to seed the sea with iron. In 1989, the oceanographer John Martin suggested that phytoplankton, which live near the surface of the ocean and pull carbon out of the air during photosynthesis, then die after about 60 days and sink to the bottom taking the carbon with them, could serve as a viable method of counteracting global warming. His theory was that if iron was pumped into the ocean, stimulating the phytoplankton to have an accelerated growth rate, they could absorb enormous amounts of carbon, then sink to the bottom of the ocean and store it away, counteracting global warming. He first published his theory in 1989 in *Nature*, calling it the iron hypothesis.

Another idea under discussion to counteract global warming is to install a pipeline to deliver iron from the coast to the ocean. The right mix of chemicals would need to be determined and the correct distance from shore would have to be calculated. It has also been suggested that wave power could help phytoplankton blooms by churning nutrient-rich waters in the deep ocean toward the surface. Another suggestion involves dumping iron dust from ships. Other scientists caution that the right chemical mix is key because phytoplankton require nitrogen, phosphorus, and other nutrients as well, so it is not simply a matter of dumping iron into the ocean. The big drawback to this idea is that there is no way to predict what side effects a massive iron infusion may have on the fragile ocean ecosystem. Another unknown is whether or not large-scale iron seeding would have enough input to be able to affect global-scale climate.

Shoot Mirrors into Space

In an attempt to deflect sunlight back into space, a third suggestion is to launch a mirror the size of Greenland and strategically position it between the Earth and the Sun. Because launching a mirror that large would be very problematic, Roger Angel, a researcher and optics expert at the University of Arizona, suggested instead launching trillions of tiny mirrors.

Angel calculated that it would take a trillion or so mirrors, each two feet (0.6 m) in diameter but only one-five-thousandth of an inch thick, to form a cloud twice the diameter of Earth. In order to stay perfectly positioned between the Earth and the Sun (which would allow about 2 percent of the sunlight to be filtered out), the mirrors would have to orbit at a region called L1, a balancing point between the Earth's and the Sun's gravitational fields.

The weight of the mirrors would be about 20 million tons (18 million metric tons). A space shuttle can only carry 25 tons (23 metric tons) at a time. This would be the equivalent of 800,000 space shuttle flights—also impractical. Even more shocking is the price tag—up to $400 trillion.

Whiten the Clouds with Wind-Powered Ships

John Latham of NCAR and Stephen Salter of the University of Edinburgh have suggested a solution based on the reflectivity of clouds. They both contend that because the tops of clouds reflect incoming solar radiation back out into space, perhaps one way to reduce the effects of global warming is to increase their reflectivity. According to Latham, "Increasing the reflective power of the clouds by just 3 percent could offset humanity's contributions to global warming; and the way to do it is to spray enormous amounts of seawater into the sky."

Both Latham and Salter suggest that a fleet of 1,500 boats could be used to spray 1,766 cubic feet (50 m³) of water droplets per second. Salter recommends that the boats be wind-powered and remotely driven so that they could be mobile, able to be located in variable locations. The ships would be powered by Flettner rotors, which are spinning cylinders that allow the boat to move perpendicularly to the wind direction. While the boats are moving, turbines being dragged through the water generate electrical energy, which goes toward blowing the droplets of water into the sky. The turbines could also be used to power the boats, if necessary, when the wind is not blowing. Brian Launder believes this is one of the most promising potential geoengineering projects. He points out that it requires very few resources—just seawater and boats. What the effects would be on the clouds, however, is not certain.

Build Fake Trees

Klaus Lackner at Columbia University has suggested another idea—physically pulling CO_2 out of the atmosphere so that it does not warm the Earth as much in the first place. In order to do this, Lackner is creating an artificial tree. His "tree" consists of panels 538 square feet (50 m^2) in size made of absorbent resin that reacts with CO_2 in the air to form a solid. When Lackner explains his trees, he compares them to a furnace filter. Just as filters pull particles out of the air, the trees pull out CO_2 from the air. When the giant panels need to be cleaned they are taken down and exposed to 113°F (45°C) steam. The chemical reaction with the steam causes the solid to release the carbon it has captured, which Lackner then consolidates as liquid CO_2.

Once the CO_2 is consolidated, it then has to be sequestered. Lackner acknowledges it can be used in greenhouses for plants to use during photosynthesis, in dry ice, or in new types of plastics and concrete that can be made with CO_2. Lackner is focusing most of his attention, however, on geological formations, specifically in porous sandstone formations under the North Sea, which he believes are viable for carbon sequestration and storage (CSS).

To date, Lackner has had problems with his trees in the Tropics because of the high humidity. He is still testing his theories in the lab and has yet to test them in the real world. He believes he may be two to three years away from having a full-scale working model. He also says that if it works, a ton of CO_2 per day may not sound like a lot, "but it is far more than your car."

Besides seeming somewhat extreme and far-fetched, a big unknown with any of these controversial geoengineering projects is that scientists do not know at this point whether or not they could shut down some of the projects once they got started. Another argument against using these extreme efforts is that geoengineering only treats the symptoms of global warming and could seriously undermine efforts to address the root cause, which is what really needs to be addressed. In addition, if scientists engineer a perceived solution to global warming, they fear that people may then feel like the threat has gone away and there is no longer a concern to reduce personal carbon emissions and imprints and that people and companies will go back to a business as usual attitude and leave the solution of global warming solely up to scientists.

Geoengineers, such as Brian Launder, do not believe that geoengineering projects should be the answer to controlling global warming for many reasons: cost, maintenance, political difficulties, and engineering difficulties to name a few. But they do believe it to be wise to research possible options so that if, in 10 or 20 years governments have still failed to take action, scientists will have feasible options ready. As Launder says, "While such geoscale interventions may be risky, the time may well come when they are accepted as less risky than doing nothing."

PRIORITIZING ADAPTATION STRATEGIES

Because global warming is already well underway and the lifetimes of GHGs can extend over 100 and more years, even if every effort possible to stop it was made immediately, people will still have to adapt for decades to come. It is already too late.

Adaptation requires the integration of climate risks into near- and long-term planning so that ecosystems and populations are able to cope with changes that can no longer be avoided. Although each geographic area is different, because of variables such as latitude, elevation, ecosystem type, presence of urban areas, humidity levels, and major atmospheric circulation systems that will require specific adaptation strategies, there are some adaptation strategies that apply to all regions and can be used in a basic planning strategy.

The UCS has identified the following eight principles that can help set priorities.

1. Monitor the changing environment: Both decision makers and resource managers need to be aware that as global warming progresses the environment will change. Therefore, it is important that the climate and the planet be monitored. Strategies may need to be adjusted over time to manage situations that may not have been planned.
2. Track indicators of vulnerability and adaptation: Decision makers need to monitor both the progress of specific adaptation strategies and the social factors that limit communities' abilities to adapt. If problems occur, adaptation strategies can then be modified so that potential outcomes are improved.
3. Take the long view: It is imperative that policy makers make decisions while planning for long-term outcomes. For

example, any investments in infrastructure, capital-intensive equipment, or irreversible land-use choices need to be made with the future in mind.
4. Consider the most vulnerable first: Climate-sensitive species, ecosystems, economic sectors, communities, and populations that are already under a considerable amount of stress for reasons other than climate change should be given high priority in policy and management decisions.
5. Build on and strengthen social networks: At the community level and within business sectors, the relationships between responsible individuals and organizations are extremely important for successful adaptation. Strong leaders have the ability to inspire organizations when times are difficult. Well-connected and respected individuals also have the ability to disseminate information more effectively that may be critical for effective adaptation.
6. Put regional assets to work: The United States has a huge wealth of scientific and technological expertise in its universities and businesses that can be used to improve the understanding of adaptation actions and challenges.
7. Improve public communication: Regular, effective communication and involvement with the public on climate change helps build the ability to successfully adapt.
8. Act swiftly to reduce emissions: Strong, immediate action to reduce emissions can slow climate change, limit the negative consequences, and give both society and ecosystems a better chance to successfully adapt to those changes that cannot be avoided.

Unless communities work together to combat global warming, it will be impossible to make the progress necessary for long-term success.

SIMPLE ACTIVITIES EVERYONE CAN DO

There are many activities to do to help cut back on personal carbon footprints. The following tables list simple actions each individual can take to help stop global warming, whether it concern transportation choices, choices at home, actions in the yard, those that make a difference in the workplace or in the community, personal choices and actions, or focus on education and public awareness.

Transportation Choices

The choice of car is one of the most important personal climate decisions someone can make. New car purchasers need to look for the best fuel economy. Each gallon of gas used contributes 25 pounds (11 kg) of GHG to the atmosphere. Better gas mileage not only reduces global warming, but also saves thousands of dollars at the pump over the life of the vehicle. Also consider new technologies like hybrid engines.

Think before driving! When someone owns more than one vehicle, they should use the less fuel-efficient one only when it will be filled with passengers. Driving a full minivan may be kinder to the environment than two midsize cars. Even better, whenever possible form a carpool or use mass transit.

With transportation accounting for more than 30 percent of U.S. CO_2 emissions, one of the best ways to reduce emissions is by riding mass transportation: buses, light rail, or subway systems. According to the American Public Transportation Association, public transit saves an estimated 1.4 billion gallons of gas annually, which translates to about 1.5 million tons (1.4 million metric tons) of CO_2. Unfortunately, 88 percent of all trips in the United States are still made by personal car.

In the airline business, several changes could help with the battle against global warming. First, if routes were adjusted so that the exit and entry points let planes fly in as straight a line as possible, that would greatly help with CO_2 emissions. As an example, last year the International Air Transport Association negotiated a more direct route from China to Europe that took an average 30 minutes off flight time, eliminating 93,476 tons (84,800 metric tons) of CO_2 annually. By unifying European airspace as a single sky it could cut fuel use up to 12 percent. Pilots also need to change the way they fly. For example, abrupt drops in altitude waste fuel, so experts are advocating continuous descents until the plane reaches the runway, where it could be towed instead of burning fuel while taxiing.

Maintain your car: An engine tune-up can improve gas mileage 4 percent; replacing a clogged air filter can increase efficiency 10 percent; and keeping tires properly inflated can improve gas mileage more than 3 percent. Although that may not seem like much, if gas mileage can be increased from even 20 to 24 MPG, a car will put 200 fewer pounds (91 kg) of CO_2 into the atmosphere each year.

Practical Solutions That Work—Getting Everyone Involved

Choices at Home
More than half the electricity in the United States comes from polluting coal-fired power plants, the single largest source of heat-trapping gas. Many states offer their customers the option of paying to enroll in renewable energy programs, such as wind energy. Although it is slightly more expensive, it helps the fight against global warming.
When household appliances need to be replaced (refrigerators, freezers, furnaces, air conditioners, and water heaters) look for the Energy Star label and purchase one of those. They may cost slightly more initially, but the energy savings will pay back the extra investment within a couple of years. Household energy savings really can make a difference. According to the UCS, if each household in the United States replaced its existing appliances with the most efficient models available, it would save approximately $15 billion in energy costs and eliminate 175 million tons (159 million metric tons) of heat-trapping gases.

(continues)

Everyone can play a part in slowing global warming by reusing grocery bags and installing compact fluorescent lightbulbs. *(Nature's Images)*

Choices at Home (continued)

Unplug a freezer. One of the most rapid ways to reduce individual global warming impact is to unplug the extra refrigerator or freezer that is rarely used (other than for holidays and parties). When these appliances are kept running with nothing in them, it adds about 10 percent to a typical family's CO_2 emissions.
Use the dishwasher only when it is full.
Wrapping a water heater in an insulated blanket can save the household about 250 pounds (113 kg) in CO_2 emissions annually. Most water heaters more than five years old are constantly losing heat and wasting energy because they lack internal insulation. If the surface feels warm to the touch, it may just need to be wrapped with an insulated blanket.
Microwave ovens reduce energy use by about two-thirds compared with conventional ovens, because they cook foods faster. Crock-pots and pressure cookers are also efficient. In addition, toaster ovens should be used if a full-size oven is not necessary.
Get a home energy audit: Take advantage of the free home energy audits that are offered by many local utility companies. Simple measures, such as installing a programmable thermostat to replace an old dial unit or sealing and insulating heating and cooling ducts, can reduce a typical family's CO_2 emissions by about 5 percent.
Lightbulbs matter! If every household in the United States replaced one regular lightbulb with an energy-saving model, it could reduce global warming pollution by more than 90 billion pounds (41 billion kg) over the life of the bulbs; which is approximately the same as taking 6.3 million cars off the road. If incandescent bulbs are replaced with efficient compact fluorescent bulbs, it will significantly cut back on heat-trapping pollution, as well as save money on electric bills and lightbulbs (the compact fluorescent bulbs are more expensive, but last much longer).
Insulate the garage, attic, and basement with natural, nontoxic materials like reclaimed blue jeans.
Protect windows from sunrays with large overhangs and double-pane glass.
Capitalize on natural cross ventilation.

Practical Solutions That Work—Getting Everyone Involved

Choices at Home *(continued)*
Hang up a clothesline: According to Cambridge University's Institute of Manufacturing, 60 percent of the energy associated with a piece of clothing is spent in washing and drying it. Over its lifetime, a T-shirt can send up to 9 pounds (4 kg) of CO_2 into the atmosphere. The solution is not to avoid doing laundry, but to wash the clothes in warm (or even cold) water instead of hot, and save up to launder a few big loads instead of many smaller ones. Use the most efficient machine available—newer ones can use as little as one-fourth the energy of older machines. When they are clean, dry clothes the natural way, by hanging them on a line rather than loading them in a dryer. By doing this, the CO_2 created by laundry can be reduced up to 90 percent.
Moving to a high-rise building also helps reduce personal carbon footprints. The smaller the living space a person occupies, the smaller the personal impact on the environment and the smaller contribution to global warming.
Open a window: Most of the 25 tons (23 metric tons) of CO_2 emissions each American is responsible for each year come from the home. Little actions can have a great impact. Opening a window instead of running the air-conditioner will get a flow of air through a home, cooling it off.
Caulk and weatherstrip all doors and windows to keep cold air from coming in during the cold winter months.
Insulate walls and ceilings.
Install low-flow showerheads.
Turn down the thermostat on the water heater.
Take care of your trash: Composting all organic waste—and recycling paper, cardboard, cans, and bottles—will help reduce the greenhouse gas emissions associated with landfills.
Try the two-degree solution: By moving the thermostat down two degrees in the winter and up 2 degrees in the summer you can save about 350 pounds (159 kg) of CO_2 emissions each year.
Switching to double-pane windows will trap more heat inside the home so that less energy needs to be used in the winter.
Switch into energy-save mode: Start using energy-saving settings on refrigerators, dishwashers, washing machines, clothes dryers, and other appliances.
Take a power-shower: showers account for two-thirds of all household water-heating costs. Cut down shower time to cut down on energy.

(continues)

Choices at Home (continued)

If remodeling a home or building a new one, incorporate energy efficiency measures into it. If possible, install a solar thermal system to help provide hot water. Consider installing a solar photovoltaic system to generate electricity.

Energy use rises and falls with the intensity of screen images, but the popular 42-inch-screen TVs with plasma technology can burn three times the power of old cathode-ray tube sets. The 42-inch LCD TVs use less energy than plasma but twice the power of an old tube TV. New Energy Star standards should help consumers seek efficient models.

Digital photo frames add just a little to each electric bill, but policymakers worry about the cumulative impact once the frames saturate the market.

Turn off and unplug whenever possible. Even when chargers or hair dryers are turned off, they still make up 5–10 percent of the electricity bill and should be unplugged when not in use. With the help of a power strip, it is easy to turn off several at once. Turn off computers and monitors that will not be used for at least 20 minutes and avoid using screen savers. Lights, TVs, stereos, and air-conditioners should be turned off when leaving a room and especially when leaving home.

In the Yard

Plant a tree: Planting a tree in the backyard is another way to combat global warming. Even better, organize a community project to plant trees on community property. In addition to storing carbon, trees planted in and around urban areas and residences can provide much-needed shade in the summer, reducing energy bills and fossil fuel use.

In *temperate* climates, do not plant deciduous trees to the south of the home. In winter, even bare branches can block the Sun from warming the home. Plant shrubs, bushes, and vines about a foot away from the wall of the home to create dead air insulating spaces. For windy areas in the winter, plant evergreen trees and shrubs close together on the northern side of the home. If snowdrifts are common, plant low shrubs to stop them from drifting up against the home.

In hot, arid climates, plant trees and shrubs that provide shade to cool roofs, walls, and windows. Make sure air-conditioning units are also shaded—this can increase efficiency by up to 10 percent. If air-conditioning is not used, make sure that summer winds are not blocked from the home by

Backyard conservation is another way to help the environment. *(ASCS)*

In the Yard *(continued)*
landscaping. Place trellises away from the wall to allow air to circulate. Vegetation that is planted too close to a home will trap summer heat and make the house feel even hotter.
In cool climates, plant dense evergreen trees and shrubs to the north and northwest of the home to protect it from cold winter winds. Evergreens can be combined with a wall, fence, or berm to lift winds over the home. If snowdrifts are common, plant low shrubs on the side of the home where the winds originate. Do not plant trees too close to the home's south side or the heating benefits of the winter Sun will be lost. Make sure not to block the Sun from south-facing windows.
In hot, humid climates plant shrubs a few feet away from the house to direct cool summer breezes toward the home. These can also provide extra shade. Plant deciduous (leafy) trees on the northeast-to-southeast and northwest-to-southwest sides of the house. Vegetation planted too close to a home will trap summer heat and make the house feel even hotter. Flowerbeds that require a lot of watering should not be planted close to the home. Plant low ground cover, including grasses, around the driveway or patio to cool these areas and prevent glare.

In the Workplace

Institute "proximate commuting." This concept works best for companies that have multiple locations in one urban area. If employees work at their company's closest business location in relationship to where they live, it not only reduces the time they spend commuting, but also helps ease rush hour traffic jams and helps with global warming.

Shut off computers. A screen saver is not an energy saver. According to the U.S. Department of Energy, 75 percent of all electricity consumed in the home is standby power used to keep electronics running when TVs, DVDs, computers, monitors, and stereos are off. The average desktop computer, not including the monitor, consumes from 60 to 250 watts a day. Compared with a machine left on all the time, a computer that is in use four hours a day and turned off the rest of the time would save about $70 a year. The carbon impact would be even greater.

Turn off all lights at quitting time. It helps battle global warming when a business has made sure each night that computers, monitors, desk lights, printers, and fax machines are turned off each night. Air-conditioners and overhead lights can be timed for turnoff.

In the Community/Education/Public Awareness

Cities can save energy and money by illuminating public spaces with LEDs, or light-emitting diodes. LEDs use 40 percent less electricity than the high-pressure sodium bulbs commonly used today. Even though they cost two to three times as much, they can go five or more years without upkeep. Traditional bulbs, on the other hand, must be replaced every 18 months.

Make policy makers aware of concern for global warming: Elected officials and business leaders need to hear from concerned citizens. Participation in organized global warming awareness groups, such as the UCS, can also help because they often have planned programs and agendas that target when specific legislation comes before Congress and can coordinate opportunities when public opinion can be heard to ensure that policy makers get the timely, accurate information they need to make informed decisions about global warming solutions.

Practical Solutions That Work—Getting Everyone Involved

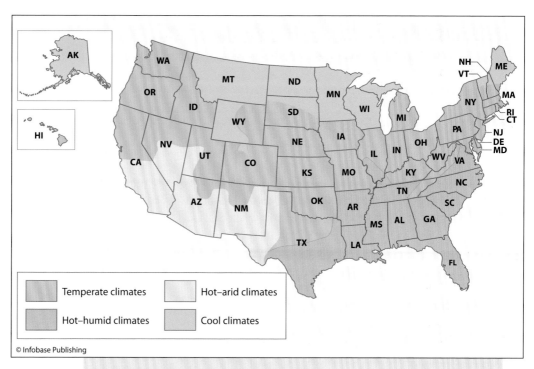

Depending on the climate zone, specific practices in landscaping can make homes more energy efficient.

In the Community/Education/ Public Awareness *(continued)*
Carbon sequestration technology is currently being looked at as a possible solution to global warming. Work still needs to be done on the technology and the price needs to come down in order to make it practical, but if these obstacles can be overcome, this technology could make a significant difference in combating global warming in the future.
Get informed and get involved. Read books and newspapers and watch films about global warming. Then talk to neighbors, coworkers, friends, family, and community groups about ways to reduce global warming.

Personal Choices and Actions
Buy sustainable wood. When buying wood products, check for labels that indicate the source of the timber. Supporting forests that are managed in a sustainable fashion makes sense for biodiversity and for the climate as well. Forests that are well managed are more likely to store carbon effectively because more trees are left standing and carbon-storing soils are less disturbed.
Cut down on paper use. Americans recycled 42 million tons of paper last year—50 percent of what they used—but still pulverized the rest. Each year, 900 million trees are cut down for the manufacture of paper products. Buying recycled paper is also recommended. It uses 60 percent less energy than virgin paper. In fact, each ton purchased saves 4,000 kWh of energy, 7,000 gallons (26,498 liters) of water, and 17 trees. In addition, for each tree saved, it has the capacity to filter up to 60 pounds (27 kg) of pollutants from the air.
Some clothing manufacturers (e.g., Patagonia) collect clothing made from Polartec and Capilene fleece to melt and recycle and make into new fabric and clothes. Interestingly, some of these materials were originally created from recycled plastic. Patagonia estimates that making polyester fiber out of recycled clothing, compared with using new polyester, will result in a 76 percent energy savings and reduce GHGs 71 percent (see www.patagonia.com/recycle for details).
Vintage clothes—high-end hand-me-downs are more ecologically sound than new clothes because buying clothes this way avoids consuming all the energy used in producing and shipping new ones, as well as all the carbon emissions that would have been emitted associated with its production. Every item of clothing has an impact on the environment. In fact, some synthetic textiles are even made with petroleum products. Cotton accounts for less than 3 percent of farmed land globally but consumes about one-fourth of the pesticides used.
Strive to become carbon neutral.
Pay your bills online. By eliminating paper trails and paying bills online, it does more than just save trees. It also helps reduce fuel consumption by the trucks and planes that transport paper checks. If every U.S. home viewed and paid its bills online, the switch would cut solid waste by 1.6 billion tons (1.5 billion metric tons) a year and cut back GHG emissions by 2.1 million tons (1.9 million metric tons) a year.

Personal Choices and Actions *(continued)*
Eat less beef. According to the U.N.'s Food and Agriculture Organization (FAO), about 18 percent of the world's GHG emissions—even more than transportation—come from the international meat industry. Much of the emissions come from the nitrous oxide in manure and the methane from the bovine's digestion.
Use reusable grocery bags. Every year, more than 500 billion plastic bags are thrown in the landfill (less than 3 percent are ever recycled). The bags are made of polyethylene and can take up to 1,000 years to biodegrade in landfills that emit harmful greenhouse gases. Today, many stores sell ecofriendly cloth grocery bags that are reusable, sturdy, and last for years. Most grocery stores also offer a five-cent rebate each time the bag is used; meaning that within a short time period, the bags have already paid for themselves many times over.
Support local farmers. By purchasing fruit, vegetables, meat, and milk produced closer to home, it helps fight global warming because the items did not have to be trucked cross-country to reach the store. Local farmers markets, greengrocers, and food co-ops can be found in *localharvest.org* by zip code. It may also be possible to join a community-supported agriculture project, which lets a consumer buy shares in a farmer's annual harvest. In return, the recipient receives a box of produce every week for a season.
Reduce, reuse, and recycle. Reduce waste by choosing reusable products instead of disposables. Buying products with minimal packaging (including the economy size when that makes sense) will help to reduce waste. Recycle paper, plastic, newspaper, glass, and aluminum cans. If there is not a recycling program available, try to start one. By recycling all household wastes, 4,800 pounds (2,177 kg) of CO_2 can be saved annually.

By following these simple suggestions, everyone can do their part to cut back on greenhouse gas emissions and make a difference for future generations. It will take every person's efforts to make the differences needed. For more suggestions on how to get involved, see the contacts listed in the Appendixes.

The Future: What Lies Ahead

Almost daily there is a news report reflecting a global warming issue: drought, floods, wildfires, hurricanes, heat waves, glaciers melting, and polar bears starving and other wildlife being threatened with extinction. Although it is not possible to forecast exactly when, where, and exactly how severe warming's impacts will be, there is enough evidence available today to understand that many of the impacts from global warming will be severe and will result in disasters with enormous economic and human costs.

Each day corrective action is delayed puts life on Earth at greater risk. What is important to realize is the climate system's inertia. Because it responds slowly, positive action taken today will not be realized for decades to come. In addition, the longer the delay, the greater the risks become and the more difficult it will be to respond effectively. Even worse, if the delay becomes too long, it may never be possible to stabilize the climate at a safe level for life to exist as it presently does. Tipping points become a serious issue—when the system tips or shifts into an entirely new state, such as the major collapse of ice sheets causing the rapid sea-level rise or massive thawing of permafrost releasing huge amounts of stored methane into the atmosphere.

Unfortunately, global warming has progressed enough that no amount of cutting back on greenhouse gas (GHG) emissions will allow some ecosystems to return to the way they once were. If emissions are cut back now on an aggressive basis, scientists at the National Aeronautics and Space Administration (NASA) believe it is still possible to avoid the worst consequences of global warming. Unlike the targeted 5 percent outlined in the Kyoto Protocol, the European Union has said that it will actually require a reduction of 60–80 percent to prevent dangerous climate change.

On the positive side, scientists do understand what the world is up against and are trying to educate the public to make the right choices, and the public does seem to be responding (although slowly) to the green movement. Solving the problem will take the concerted effort of everyone. There will have to be change in the future design of buildings, transportation, energy systems, leadership, innovation, and investments from governments and businesses. Both public and individual commitments are critical in order to achieve success.

This chapter looks at the future and how several leading scientists expect the world to become under the influence of increased global warming. It then looks at the predicted winners and losers in the future as the Earth continues to heat up. In conclusion, it looks at what new technology is on the horizon to help manage for a better tomorrow.

A LOOK TOWARD THE FUTURE

Based on several emission scenarios run by the Intergovernmental Panel on Climate Change (IPCC) and NASA, global temperature is projected to increase by approximately 0.3°F (0.2°C) per decade for the next two decades. Even if GHGs were kept steady at 2000 levels, because of the inertia of the oceans—the long time it takes them to store and release heat—there is already a suggested warming in the pipeline of 0.2°F (0.1°C) per decade.

If GHG emissions continue at the current rate, or become even greater, climate models suggest that changes in the global climate system during this century will be even larger than those that were observed during the 20th century. Another critical factor is the warmer it gets, the less CO_2 the land and ocean are physically able to store. This means that any increasing concentrations in CO_2 will remain in the atmosphere. At this point, the IPCC projects that from now to 2090, the global average surface air warming will most likely range from 1.8–10.7°F (1.1–6.4°C). The ranges are attributed to the differences in the models and energy-use scenarios used.

Global average sea level is projected to rise by 7–23 inches (18–59 cm) by 2099. Scientists caution, however, that models do not include uncertainties about some climate mechanisms because there is still a lack of knowledge. For example, one of the key uncertainties is ice flow from Greenland and Antarctica. There are still mechanisms that control the flow and dynamics of the ice that scientists are trying to understand. If the speed of future ice flow increases, it will affect future scenarios that may not be accounted for at this point.

The geographical patterns in climate changes are expected to remain similar to those observed over the past several decades. The areas expected to be affected with warming the most are the high northern latitudes (polar region) and over the Earth's landmasses. The least amount of warming is expected over the Southern Ocean and parts of the north Atlantic Ocean.

There are several other predicted changes that will also occur by 2099:

- Snow cover and sea ice will continue to shrink, endangering polar bears and other arctic animals, and permafrost will melt, releasing methane.
- As CO_2 increases in the atmosphere, the oceans will become more acidic.

- There will be increasing frequent heat waves, hot extremes, and heavy precipitation events.
- More intense and frequent hurricanes are likely to occur.
- There will be a moving of extratropical storm tracts toward the poles, with changes in wind, precipitation, and temperature patterns.
- A greater amount of precipitation in high latitudes and less rain in most subtropical land regions is expected.
- A slowing of the Atlantic Ocean Circulation (a major transporter of global heat) will likely occur.

IPCC scientists believe that warming and sea-level rise will continue for centuries even if GHG emissions were to become stabilized because of the long timescales associated with climate processes and feedbacks. One significant uncertainty is that global warming is expected to affect the Earth's carbon cycle, but exactly how and by how much is not known at this point. Even if GHG emissions stopped and were stabilized by 2100, the Earth's atmosphere would still warm by approximately 0.8°F (0.5°C) by 2200. The thermal expansion of the oceans alone would cause an increase of 12–31 inches (30–80 cm) of global sea-level rise by 2030. The ocean would continue to warm for many centuries after that.

Greenland's ice sheet is projected to keep melting and also cause sea levels to rise after 2100. NASA scientists say that if it were to continue to melt for thousands of years until it completely melted, it would cause global sea levels to rise about 23 feet (7 m). It is not well understood yet what exactly will happen to the Antarctic ice sheets. Their vulnerability could increase through dynamical processes related to ice flow—these details are not included in current models but have been observed in the field. Future models will need to include field observations as they occur and become better understood. Their melting could also add to the global sea level. Current global models so far suggest that the Antarctic ice sheet will stay too cold throughout this century for large-scale surface melting. Some models project that the ice sheet could even increase in mass due to increased snowfall events. Scientists at NASA's Goddard Institute for

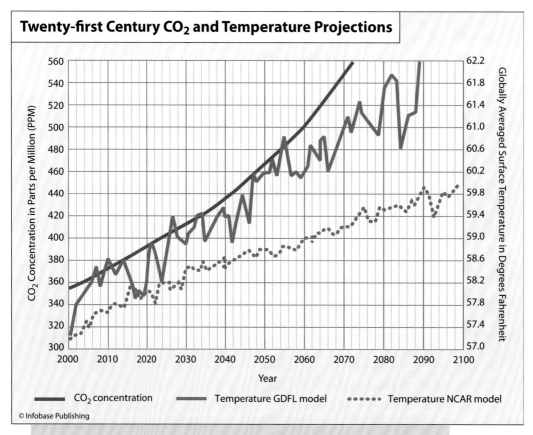

CO_2 is projected to continue to rise throughout this century, and temperatures, according to two different climate models, will continue to climb under the influence of global warming.

Space Studies (GISS) believe that past, present, and future emissions of CO_2 will contribute to warming and sea-level rise for more than the next thousand years because of the length of time the greenhouse gases remain in the atmosphere.

In a *LiveScience* time line published on April 19, 2007, the following major changes were predicted for the environment if global warming continued. With predictions like these, it is not hard to see why scientists and environmentalists alike are lobbying for proactive action to take place now. Prolonging action any further is only going to make future adaptation more difficult.

Future Changes in the Earth's Environment under the Effects of Continued Global Warming

2020
- Flash floods will increase across Europe (IPCC).
- Less rainfall could reduce agricultural yields up to 50 percent in some areas of the world (IPCC).
- World population will reach 7.6 billion people (U.S. Census Bureau).

2030
- Up to 18 percent of the world's coral reefs will probably die because of climate change and other environmental stresses (IPCC).
- World population will reach 8.3 billion people. (U.S. Census Bureau).
- Warming temperatures will cause temperate glaciers on equatorial mountains in Africa to disappear (IPCC).

2040
- The Arctic Sea could be ice free in the summer and winter ice depth may shrink drastically (IPCC).

2050
- Small alpine glaciers will very likely disappear completely and large glaciers will shrink by 30 to 70 percent (IPCC).
- As biodiversity hot spots are more threatened, one-fourth of the world's plant and vertebrate animal species could face extinction (IPCC).

2070
- As glaciers disappear and areas affected by drought increase, electricity production for the world's existing hydropower stations will decrease. The hardest hit will be Europe, where hydropower potential is expected to decline on average by 6 percent; around the Mediterranean, the decrease could be up to 50 percent (IPCC).
- Warmer and drier conditions will lead to more frequent drought, more wildfires, and more frequent heat waves, especially in Mediterranean regions (IPCC).

2080
- Some parts of the world will be flooded. Up to 20 percent of the world's population lives in river basins that will be hit with increasing flood hazards. Up to 100 million people could experience coastal flooding each year. The most vulnerable are the densely populated low-lying areas (IPCC).
- Between 1.1 and 3.2 billion people will experience water shortages and up to 600 million will go hungry (IPCC).

(continues)

> **Future Changes in the Earth's Environment under the Effects of Continued Global Warming**
> *(continued)*
>
> **2100**
> - Atmospheric CO_2 levels will be higher than they have been for the last 650,000 years (IPCC).
> - Ocean pH levels will very likely decrease by as much as 0.5 pH units—the lowest it has been in the last 20 million years (IPCC).
> - Thawing permafrost will make Earth's land area a new source of carbon emissions—it will emit more CO_2 into the atmosphere than it absorbs (IPCC).
> - New climate zones will appear on up to 39 percent of Earth's land surface (IPCC).
> - One-fourth of all plant and land animal species could become extinct (IPCC).

WINNERS AND LOSERS

In the United States, the U.S. Global Change Research Program operates the U.S. National Assessment of the Potential Consequences of Climate Variability and Change (National Assessment). It breaks the United States into regional geographic sections (e.g., Pacific Northwest, Southeast, West, Midwest, Great Plains, Alaska, and so forth), and generates reports for each region detailing climate impacts and practical methods of adaptation.

The reports are backed by both scientists and policy makers in the hope that constructive progress will be made in scientific understanding and social action. The way the program works, the National Assessment currently consists of 16 ongoing regional projects. For each of the regional studies, teams of scientists, resource planners, and other involved parties meet to assess the region's most critical vulnerabilities in areas such as agricultural productivity, coastal areas, water resources, forests, and human health. In addition to looking at potential impacts, the teams also work together to identify possible strategies that can be used to adapt and respond to climate change. The overall goal of the project is to help those in the United States prepare for future climate change.

According to Michael MacCracken, who heads the national office, "The goal of the assessment is to provide the information for communities as well as activities to prepare and adapt to the changes in climate that are starting to emerge."

A *New York Times* article of April 2, 2007, outlined which countries will be hit the hardest as climate change progresses. There will be what they refer to as winners and losers. It is known that the industrialized nations are the largest producers of GHGs. In general, it is the industrialized countries that are also the best equipped to deal with the effects of global warming and to mitigate by financing adaptive measures. Unfortunately, it is the poorer nations that lie closer to the Tropics—even though they have not contributed to the GHG emission problem as significantly—that will be dealt the majority of the worst side effects, such as drought, crop failure, heat waves, flooding, and sea-level rise.

The *Times* article mentioned several geographic areas worldwide that are already in the process of adapting to climate change. In Shishmaref, Alaska, for example, on a low-lying island, the town is in the process of relocating because the island is already being eroded by changes in sea level. The estimated costs to relocate are estimated at $180 million. The shoreline has receded three to five feet (0.9–1.5 m) per year and is especially vulnerable when tidal high water is combined with intense wave action of the Chukchi Sea during storms. The community is relocating to an area on the mainland that is accessible to the sea and will provide the community with the subsistence lifestyle they are used to, allowing them to hold on to their tribal culture.

The U.S. corn belt is genetically modifying crop varieties that are designed for drought and pest resistance so that farmers will be able to sustain their yields in the hotter, drier years to come. London is currently making improvements to their flood protection infrastructure on the Thames River to guard against flooding events as the climate warms. On Sylt Island in Germany, a pilot project is underway to build more resilient dykes out of rocks that are precoated with flexible polyurethane. This keeps the dike from being weathered by the North Sea by both absorbing the force of the breaking waves and slowing down the water masses.

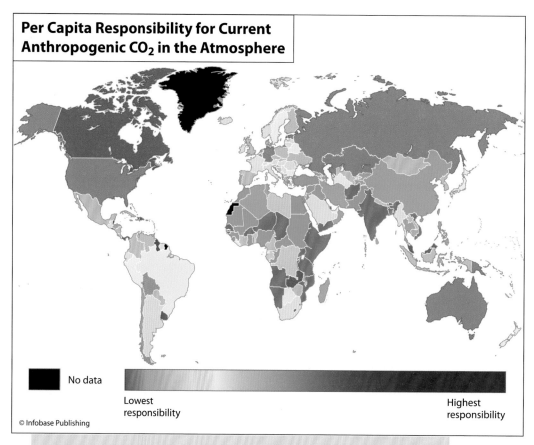

This map shows the per capita responsibility for GHGs worldwide. In many cases, the largest offenders are often the most wealthy, industrialized nations that are likely to encounter the least in losses overall because they have many of the resources necessary to mitigate the negative effects. Unfortunately, the countries that are likely to encounter the greatest losses are the undeveloped countries and those located close to sea level that have not contributed significantly to the global warming problem. *(Source: World Resources Institute)*

In Andermatt, Switzerland, one ski resort has had to construct a ramp each ski season in order to gain access to a steadily receding glacier. The ramp has now been covered with a reflective cover to protect it from melting. Venice, Italy, which is extremely prone to flooding from sea-level rise, is constructing floodgates to protect the city's infrastructure during extremely high tides. In Northern China in a very dry region, a project is under construction to divert water hundreds of miles from the Yangtze River in the south. In Perth, Australia, they have finished construction on a desalination plant to serve as a backup source of water to offset shrinking natural supplies as a result of prolonged drought conditions. As global warming continues, locations will have to continue to adapt to changing conditions as they arise.

NEW TECHNOLOGIES

In an April 2008 *Scientific American* article, Jeffrey D. Sachs, head of the Earth Institute at Columbia University, said that, "Even with a cutback in wasteful energy spending, our current technologies cannot support both a decline in carbon dioxide emissions and an expanding global economy. If we try to restrain emissions without a fundamentally new set of technologies, we will end up stifling economic growth, including the development prospects for billions of people."

What Mr. Sachs says is needed is a huge investment of resources and effort into new technologies that are low carbon and this will not happen with the kind of effort toward research that has occurred so far. It will require the serious, dedicated involvement of determined government leadership and resources; a program so intense and focused, he refers to it as a "Manhattan-like Project."

As researchers learn more about global warming and gain a better understanding of the complex interactions of the climate system, this knowledge coupled with technology should lead to even better solutions. Over the past 30 years, computing power has increased by a factor of 1 million. Models today are becoming much more complex and realistic. As a better understanding is reached of the nature of feedbacks from the carbon cycle and their constraints on the climate response, models are becoming much more sophisticated. New "petascale" computer models depicting detailed climate dynamics are now building the foundation for the next generation of complex climate models. New advanced computing abilities will help climatologists better understand the links between weather and climate.

This new technology is being developed by researchers at the University of Miami Rosenstiel School of Marine and Atmospheric Science

(RSMAS), the National Center for Atmospheric Research (NCAR) in Boulder, Colorado, the Center for Ocean-Land-Atmospheric Studies (COLA) in Calverton, Maryland, and the University of California at Berkeley. They are using a $1.4 million award from the National Science Foundation (NSF) to generate the new models.

The scientists at these institutions say that the development of powerful supercomputers capable of analyzing decades of data in the blink of an eye marks a technological milestone capable of bringing comprehensive changes to science. The speed of supercomputing is measured in how many calculations can be performed in a given second. Petascale computers can make 1,000,000,000,000,000 calculations per second, an enormous amount of calculations even when comparing it to an advanced supercomputer. Because of the "peta's" capabilities, this represents a breakthrough and a golden opportunity for climatologists to advance Earth science system science and help to improve the quality of life on the planet.

Jay Fein, NSF program director, says, "The limiting factor to more reliable climate predictions at higher resolution is not scientific ideas, but the computational capacity to implement those ideas. This project is an important step forward in providing the most useful scientifically based climate change information to society for adapting to climate change."

One thing researchers have learned recently through modeling is that climate cannot be predicted independently of weather. They have discovered that weather has a profound impact on climate. Now that they have discovered this, they expect to be able to greatly improve weather and climate predictions and climate change projections. In addition, with the increase in computing capabilities, one of the team members—Ben Kirtman, a meteorologist at RSMAS—has developed a new weather and climate modeling strategy which he calls "interactive ensembles," which is designed to isolate the interactions between weather and climate.

The interactive ensembles for weather and climate modeling are currently being applied to one of the United States' main climate change models—NCAR's Community Climate System Model (CCSM), the current operational model used by NOAA's climate forecast system (CFS).

Continual warming temperatures will eventually affect every person on the Earth. *(Nature's Images)*

The CCSM is also a model used by hundreds of researchers and is also one of the climate models that was used in the Nobel Prize–winning IPCC assessments.

The research currently being done serves as a pilot program for the implementation of even more complicated computational systems, which, today, still remain a scientific and engineering challenge.

According to Kirtman, "This marks the first time that we will have the computational resources available to address these scientific challenges in a comprehensive manner. The information from this project will serve as a cornerstone for petascale computing in our field and help to advance the study of the interactions between weather and climate phenomena on a global scale." Models will continue to play an important role in the future, and, as more of the interactions between the atmosphere, geosphere, biosphere, and hydrosphere are understood along with the carbon cycle, greater insight will become available as to how more efficiently to combat global warming.

THE FINAL CHOICE

The defining moment will be when people realize that it is a personal decision must be made jointly by every individual on Earth. And that choice will be as individual as the person himself. It will be a compilation of personal values, beliefs, character, and goals for the future. And each individual's choice will count; each will have equal weight in this war against time.

In the end, each person will have to study the issues and make up their own mind. Each person will have to assess how their actions may affect the lives of their children, grandchildren, and future generations.

> *A plant takes from the soil only what it needs.*
> *In the same way, we too should only take from the*
> *Earth what we need to flourish.*
> —Chiara Lubich to young people

The solution to global warming is tied to each individual on this Earth. Ultimately, the solution boils down to one question: "How much are you willing to sacrifice to do your part?"

APPENDIX A

UNFCCC MEMBER NATIONS

Afghanistan
Albania
Algeria
Angola
Antigua and Barbuda
Argentina
Armenia
Australia
Austria
Azerbaijan
Bahamas
Bahrain
Bangladesh
Barbados
Belarus
Belgium
Belize
Benin
Bhutan
Bolivia
Bosnia and Herzegovina
Botswana
Brazil
Brunei
Bulgaria
Burkina Faso
Burundi
Cambodia
Cameroon

Canada
Cape Verde
Central African Republic
Chad
Chile
China
Colombia
Comoros
Congo, Democratic Republic of the
Congo, Republic of the
Cook Islands
Costa Rica
Côte d'Ivoire
Croatia
Cuba
Cyprus
Czech Republic
Denmark
Djibouti
Dominica
Dominican Republic
Ecuador
Egypt
El Salvador
Equatorial Guinea
Eritrea
Estonia
Ethiopia

European Union
Fiji
Finland
France
Gabon
Gambia
Georgia
Germany
Ghana
Greece
Grenada
Guatemala
Guinea
Guinea-Bissau
Guyana
Haiti
Honduras
Hungary
Iceland
India
Indonesia
Iran
Ireland
Israel
Italy
Jamaica
Japan
Jordan
Kazakhstan
Kenya
Kiribati
Korea, North
Korea, South
Kuwait
Kyrgyzstan
Laos
Latvia
Lebanon
Lesotho
Liberia
Libya
Liechtenstein
Lithuania
Luxembourg
Macedonia, Republic of
Madagascar
Malawi
Malaysia
Maldives
Mali
Malta
Marshall Islands
Mauritania
Mauritius
Mexico
Micronesia, Federated States of
Moldova
Monaco
Mongolia
Montenegro
Morocco
Mozambique
Myanmar
Namibia
Nauru
Nepal
Netherlands
New Zealand
Nicaragua
Niger
Nigeria
Niue

Appendix A

Norway
Oman
Pakistan
Palau
Panama
Papua New Guinea
Paraguay
Peru
Philippines
Poland
Portugal
Qatar
Romania
Russia
Rwanda
Saint Kitts and Nevis
Saint Lucia
Saint Vincent and the Grenadines
Samoa
San Marino
Sao Tome and Principe
Saudi Arabia
Senegal
Serbia
Seychelles
Sierra Leone
Singapore
Slovakia
Slovenia
Solomon Islands
South Africa
Spain

Sri Lanka
Sudan
Suriname
Swaziland
Sweden
Switzerland
Syria
Tajikistan
Tanzania
Thailand
Timor-Leste
Togo
Tonga
Trinidad and Tobago
Tunisia
Turkey
Turkmenistan
Tuvalu
Uganda
Ukraine
United Arab Emirates
United Kingdom
United States
Uruguay
Uzbekistan
Vanuatu
Venezuela
Vietnam
Yemen
Zambia
Zimbabwe

APPENDIX B

HOW TO TAKE ACTION NOW

The following is a list of several notable organizations and personnel that provide research, information, and ways to get personally involved in fighting global warming. The Web sites were all accessed for availability as of May 15, 2009.

AL GORE

With the humor and humanity exuded in *An Inconvenient Truth,* Al Gore spells out 15 ways that individuals can address climate change immediately, from buying a hybrid to inventing a new, hotter "brand name" for global warming. URL: http://www.ted.com/index.php/talks/lang/eng/al_gore_on_averting_climate_crisis.html.

BEYOND KYOTO

An essay that argues for small steps to reduce carbon dioxide emissions today that can make a big difference down the road. URL: http://www.foreignaffairs.com/articles/59916/john-browne/beyond-kyoto.

CARBON CALCULATOR FOR YOUR CARBON USE

Everyone contributes to global warming every day. The CO_2 everyone produces by driving the car and leaving the lights on adds up quickly. It is surprising how much CO_2 each person emits each year. Calculate your personal impact and learn how to take action to reduce or even eliminate individual emissions of CO_2. URL: http://www.climatecrisis.net/takeaction/carboncalculator/.

CENTER FOR BIOLOGICAL DIVERSITY

The center believes that the welfare of human beings is deeply linked to nature—to the existence in the world of a vast diversity of wild animals and plants. Because diversity has intrinsic value and because its loss impoverishes society, they work to secure a future for all species, great and small, hovering on the brink of extinction. They do so through science, law, and creative media, with a focus on protecting the lands, waters, and climate that each species needs to survive. URL: http://www.biologicaldiversity.org/index.html.

CITIES FOR CLIMATE PROTECTION

The International Council for Local Environmental Initiatives enlists cities to adopt policies and implement measures to achieve quantifiable reductions in local GHGs, improve air quality, and enhance the urban quality of life. URL: http://www.iclei.org/index.php?ie=800.

CLEAN AIR COOL PLANET'S CAMPUS CLIMATE ACTION TOOLKIT FOR COLLEGES AND UNIVERSITIES

Offers helpful information on the steps that can be taken to make educational institutions more climate friendly. The site includes guidance for every aspect of campus climate action along with hyperlinks to technical resources and examples/case studies that will help people understand, plan, and execute or implement a climate action plan's various elements. URL: www.cleanair-coolplanet.org/toolkit/.

CLIMATE ACTION REGISTRY REPORT ON-LINE TOOL (CARROT)

Helping companies monitor emission reduction goals. It features an online GHG reporting and calculation tool. URL: https://www.climateregistry.org.

CLIMATE CAMP 2009

A coalition of individuals and groups that weaves four key themes—education, direct action, sustainable living, and building a movement—to tackle climate change both resisting climate crimes and developing sustainable solutions. URL: http://climatecamp.org.uk/?q=node/468.

CLIMATE HOT MAP

An interactive map based on IPCC report data illustrating the most vulnerable areas worldwide under the influence and progression of global warming. URL: http://www.climatehotmap.org/.

CONVERGENCE FOR CLIMATE ACTION—SUMMER 2008

U.S. member of international grassroots movement that plans to work with and support communities on the front lines of the climate struggle and take direct action against the fossil fuel empire. URL: http://www.climateconvergence.org/.

EARTH INSTITUTE

The Earth Institute's goal is to help achieve sustainable development primarily by expanding the world's understanding of Earth as one integrated system. They work toward this goal through scientific research, education, and the practical application of research for solving real-world challenges. With 850 scientists, postdoctoral fellows, and students working in and across more than 20 Columbia University research centers, the Earth Institute is helping to advance nine interconnected global issues: climate and society, water, energy, poverty, ecosystems, public health, food and nutrition, and hazards and urbanization. URL: http://www.earth.columbia.edu/articles/view/2124.

EARTHJUSTICE

EarthJustice is a nonprofit public interest law firm dedicated to protecting the magnificent places, natural resources, and wildlife of this Earth and to defending the rights of all people to a healthy environment. They bring about far-reaching change by enforcing and strengthening environmental laws on behalf of hundreds of organizations, coalitions, and communities. URL: http://www.earthjustice.org/.

ENVIRONMENTAL DEFENSE FUND

The Environmental Defense is a not-for-profit environmental advocacy group with four main goals: (1) stabilizing the Earth's climate, (2) safeguarding the world's oceans, (3) protecting human health, and (4) defending and restoring biodiversity. They start with rigorous science, and then work directly with businesses, government, and communities. Together, they create lasting solutions to the most serious environmental problems. One of their top priorities is to pass national legislation that caps global warming pollution and creates a flexible emissions trading market so that it will open the door to a "green technology revolution." URL: http://www.edf.org/home.cfm.

GLOBAL GREEN

Founded in 1993 by activist and philanthropist Diane Meyer Simon, Global Green is the American Arm of Green Cross International (GCI), which was created by President Mikhail S. Gorbachev to foster a global value shift toward a sustainable and secure future by reconnecting humanity with the environment. Global Green is working to address some of the greatest challenges facing humanity. In the United States their work is primarily focused on stemming global climate change by creating green buildings and cities. URL: http://globalgreen.org/.

GREENPEACE

Their core values are reflected in their environmental campaign work: "We 'bear witness' to environmental destruction in a peaceful, nonviolent manner. We use nonviolent confrontation to raise the level and quality of public debate. In exposing threats to the environment and finding solutions we have no permanent allies or adversaries. We ensure our financial independence from political or commercial interests. We seek solutions for, and promote open, informed debate about society's environmental choices." URL: http://www.greenpeace.org/usa/.

INTERNATIONAL COUNCIL ON LOCAL ENVIRONMENTAL INITIATIVES (ICLEI)

Technical support and toolkits for participants in the Cities for Climate Protection campaign. URL: www.iclei.org.

NATIONAL GEOGRAPHIC

Since 1888, they have traveled the Earth, sharing its amazing stories with each new generation. *National Geographic*'s mission programs support critical expeditions and scientific fieldwork, encourage geographic education for students, promote natural and cultural conservation, and inspire audiences through new media, vibrant exhibitions, and live events. URL: http://news.nationalgeographic.com.

NATURAL RESOURCES DEFENSE COUNCIL

NRDC uses law, science, and the support of more than 500,000 members nationwide to protect the planet's wildlife and wild places and to ensure a safe and healthy environment for all living things. NRDC's "Turn Up the Heat: Fight Global Warming" campaign empowers people concerned about global warming to turn up the heat on politicians and business leaders to act now to put solutions to global warming in place. NRDC is the nation's most effective environmental action group, combining the grassroots power of 1.2 million members and online activists with the courtroom clout and expertise of more than 350 lawyers, scientists and other professionals. URL: http://www.nrdc.org/.

NORTHWEST EARTH INSTITUTE

The Northwest Earth Institute is recognized as a national leader in the development of innovative programs that empower individuals and organizations to transform culture toward a sustainable and enriching future. They offer an educational course on global warming. URL: http://www.nwei.org/.

PEW CLIMATE GROUP

Business leadership group that offers strategies to fight global warming. Offers lots of educational resources. URL: www.pewclimate.org.

SIERRA CLUB

"Protecting the environment . . . for our families, for our future." The Sierra Club has been devoted to preserving nature's miracles for over 100 years. This map builds on work undertaken by the Sierra Club, which published a report on global warming and extreme weather in the United States in August 1998. URL: http://www.sierraclub.org/.

UNION OF CONCERNED SCIENTISTS

UCS is an independent nonprofit alliance of 50,000 concerned citizens and scientists across the country. They augment rigorous scientific analysis with innovative thinking and committed citizen advocacy to build a cleaner, healthier environment and a safer world. URL: http://www.ucsusa.org/.

USA'S ENVIRONMENTAL PROTECTION AGENCY'S STATE ACTION PLANS

This Web site offers information on current climatic changes; emissions and concentrations of greenhouse gases; local climate changes; future climate changes; effects on human health; and the status of climate change impact to water resources, agriculture, forests, and ecosystems for each individual state in the United States. URL: http://yosemite.epa.gov/oar/globalwarming.nsf/content/ImpactsStateImpacts.html.

U.S. CONGRESS

A resource to find out who are your state senators and representatives in Washington, D.C. URL: http://clerk.house.gov/member_info/index.html.

U.S. GREEN BUILDING COUNCIL'S LEADERSHIP IN ENERGY AND ENVIRONMENTAL DESIGN (LEED)

This Web site offers criteria for efficient building construction and renovation, being adopted widely by the U.S. military and others. URL: www.usgbc.org.

U.S. PUBLIC INTEREST RESEARCH GROUP

The state PIRGs created the U.S. Public Interest Research Group (U.S. PIRG) in 1983 to act as watchdog for the public interest in the nation's capital, as much as PIRGs have worked to safeguard the public interest in state capitals since 1971. URL: http://www.uspirg.org/.

WILDERNESS SOCIETY

The society's chief goal is to protect America's wilderness, not as a relic of the nation's past, but as a thriving ecological community that is central to life itself. To meet their goals, they use science and collaboration with communities and conservation groups to bring about sensible policies and positive change in land conservation. URL: http://wilderness.org/.

WORLD RESOURCES INSTITUTE

WRI provides information, ideas, and solutions to global environmental problems. Their mission is to move human society to live in ways that protect Earth's environment for current and future generations. Their program meets global challenges by using knowledge to catalyze public and private action. URL: http://www.wri.org/.

WORLD WILDLIFE FUND

World Wildlife Fund (WWF) is dedicated to protecting the world's wildlife and wild lands. WWF directs its conservation efforts toward three global goals: protecting endangered spaces, saving endangered species, and addressing global threats. URL: http://www.worldwildlife.org/.

CHRONOLOGY

ca. 1400–1850 Little Ice Age covers the Earth with record cold, large glaciers, and snow. There is widespread disease, starvation, and death.

1800–70 The levels of CO_2 in the atmosphere are 290 ppm.

1824 Jean-Baptiste Joseph Fourier, a French mathematician and physicist, calculates that the Earth would be much colder without its protective atmosphere.

1827 Jean-Baptiste Joseph Fourier presents his theory about the Earth's warming. At this time many believe warming is a positive thing.

1859 John Tyndall, an Irish physicist, discovers that some gases exist in the atmosphere that block infrared radiation. He presents the concept that changes in the concentration of atmospheric gases could cause the climate to change.

1894 Beginning of the industrial pollution of the environment.

1913–14 Svante Arrhenius discovers the greenhouse effect and predicts that the Earth's atmosphere will continue to warm. He predicts that the atmosphere will not reach dangerous levels for thousands of years, so his theory is not received with any urgency.

1920–25 Texas and the Persian Gulf bring productive oil wells into operation, which begins the world's dependency on a relatively inexpensive form of energy.

1934 The worst dust storm of the dust bowl occurs in the United States on what historians would later call Black Sunday. Dust storms are a product of drought and soil erosion.

1945 The U.S. Office of Naval Research begins supporting many fields of science, including those that deal with climate change issues.

1949–50 Guy S. Callendar, a British steam engineer and inventor, propounds the theory that the greenhouse effect is linked to human actions and will cause problems. No one takes him too seriously, but scientists do begin to develop new ways to measure climate.

1950–70 Technological developments enable increased awareness about global warming and the enhanced greenhouse effect. Studies confirm a steadily rising CO_2 level. The public begins to notice and becomes concerned with air pollution issues.

1958 U.S. scientist Charles David Keeling of the Scripps Institution of Oceanography detects a yearly rise in atmospheric CO_2. He begins collecting continuous CO_2 readings at an observatory on Mauna Loa, Hawaii. The results became known as the famous Keeling Curve.

1963 Studies show that water vapor plays a significant part in making the climate sensitive to changes in CO_2 levels.

1968 Studies reveal the potential collapse of the Antarctic ice sheet, which would raise sea levels to dangerous heights, causing damage to places worldwide.

1972 Studies with ice cores reveal large climate shifts in the past.

1974 Significant drought and other unusual weather phenomenon over the past two years cause increased concern about climate change not only among scientists but with the public as a whole.

1976 Deforestation and other impacts on the ecosystem start to receive attention as major issues in the future of the world's climate.

1977 The scientific community begins focusing on global warming as a serious threat needing to be addressed within the next century.

1979 The World Climate Research Programme is launched to coordinate international research on global warming and climate change.

1982	Greenland ice cores show significant temperature oscillations over the past century.
1983	The greenhouse effect and related issues get pushed into the political arena through reports from the U.S. National Academy of Sciences and the Environmental Protection Agency.
1984–90	The media begins to make global warming and its enhanced greenhouse effect a common topic among Americans. Many critics emerge.
1987	An ice core from Antarctica analyzed by French and Russian scientists reveals an extremely close correlation between CO_2 and temperature going back more than 100,000 years.
1988	The United Nations set up a scientific authority to review the evidence on global warming. It is called the Intergovernmental Panel on Climate Change (IPCC) and consists of 2,500 scientists from countries around the world.
1989	The first IPCC report says that levels of human-made greenhouse gases are steadily increasing in the atmosphere and predicts that they will cause global warming.
1990	An appeal signed by 49 Nobel prizewinners and 700 members of the National Academy of Sciences states, "There is broad agreement within the scientific community that amplification of the Earth's natural greenhouse effect by the buildup of various gases introduced by human activity has the potential to produce dramatic changes in climate . . . Only by taking action now can we insure that future generations will not be put at risk."
1992	The United Nations Conference on Environment and Development (UNCED), known informally as the Earth Summit, begins on June 3 in Rio de Janeiro, Brazil. It results in the United Nations Framework Convention on Climate Change, Agenda 21, the Rio Declaration on Environment and Development Statement of Forest Principles, and the United Nations Convention on Biological Diversity.

1993 Greenland ice cores suggest that significant climate change can occur within one decade.

1995 The second IPCC report is issued and concludes there is a human-caused component to the greenhouse effect warming. The consensus is that serious warming is likely in the coming century. Reports on the breaking up of Antarctic ice sheets and other signs of warming in the polar regions are now beginning to catch the public's attention.

1997 The third conference of the parties to the Framework Convention on Climate Change is held in Kyoto, Japan. Adopted on December 11, a document called the Kyoto Protocol commits its signatories to reduce emissions of greenhouse gases.

2000 Climatologists label the 1990s the hottest decade on record.

2001 The IPPC's third report states that the evidence for anthropogenic global warming is incontrovertible, but that its effects on climate are still difficult to pin down. President Bush declares scientific uncertainty too great to justify Kyoto Protocol's targets.

The United States Global Change Research Program releases the findings of the National Assessment of the Potential Consequences of Climate Variability and Change. The assessment finds that temperatures in the United States will rise by 5 to 9°F (3–5°C) over the next century and predicts increases in both very wet (flooding) and very dry (drought) conditions. Many ecosystems are vulnerable to climate change. Water supply for human consumption and irrigation is at risk due to increased probability of drought, reduced snow pack, and increased risk of flooding. Sea-level rise and storm surges will most likely damage coastal infrastructure.

2002 Second hottest year on record.

Heavy rains cause disastrous flooding in Central Europe leading to more than 100 deaths and more than $30 billion in damage. Extreme drought in many parts of the world (Africa, India,

Australia, and the United States) results in thousands of deaths and significant crop damage. President Bush calls for 10 more years of research on climate change to clear up remaining uncertainties and proposes only voluntary measures to mitigate climate change until 2012.

2003 U.S. senators John McCain and Joseph Lieberman introduce a bipartisan bill to reduce emissions of greenhouse gases nationwide via a greenhouse gas emission cap and trade program.

Scientific observations raise concern that the collapse of ice sheets in Antarctica and Greenland can raise sea levels faster than previously thought.

A deadly summer heat wave in Europe convinces many in Europe of the urgency of controlling global warming but does not equally capture the attention of those living in the United States.

International Energy Agency (IEA) identifies China as the world's second largest carbon emitter because of their increased use of fossil fuels.

The level of CO_2 in the atmosphere reaches 382 ppm.

2004 Books and movies feature global warming.

2005 Kyoto Protocol takes effect on February 16. In addition, global warming is a topic at the G8 summit in Gleneagles, Scotland, where country leaders in attendance recognize climate change as a serious, long-term challenge.

Hurricane Katrina forces the U.S. public to face the issue of global warming.

2006 Former U.S. vice president Al Gore's *An Inconvenient Truth* draws attention to global warming in the United States.

Sir Nicholas Stern, former World Bank economist, reports that global warming will cost up to 20 percent of worldwide gross domestic product if nothing is done about it now.

2007 IPCC's fourth assessment report says glacial shrinkage, ice loss, and permafrost retreat are all signs that climate change is underway now. They predict a higher risk of drought, floods, and more powerful storms during the next 100 years.

Al Gore and the IPCC share the Nobel Peace Prize for their efforts to bring the critical issues of global warming to the world's attention.

2008 The price of oil reached and surpassed $100 per barrel, leaving some countries paying more than $10 per gallon.

Energy Star appliance sales have nearly doubled. Energy Star is a U.S. government-backed program helping businesses and individuals protect the environment through superior energy efficiency.

U.S. wind energy capacity reaches 10,000 megawatts, which is enough to power 2.5 million homes.

2009 President Obama takes office and vows to address the issue of global warming and climate change by allowing individual states to move forward in controlling greenhouse gas emissions. As a result, American automakers can prepare for the future and build cars of tomorrow and reduce the country's dependence on foreign oil. Perhaps these measures will help restore national security and the health of the planet, and the U.S. government will no longer ignore the scientific facts.

The year 2009 will be a crucial year in the effort to address climate change. The meeting on December 7–18 in Copenhagen, Denmark, of the UN Climate Change Conference promises to shape an effective response to climate change. The snapping of an ice bridge in April 2009 linking the Wilkins Ice Shelf (the size of Jamaica) to Antarctic islands could cause the ice shelf to break away, the latest indication that there is no time to lose in addressing global warming.

The EPA took a major step on December 7, 2009, by declaring carbon dioxide a dangerous pollutant, allowing the Obama administration to regulate the tailpipe and smokestack emissions that add to global warming.

GLOSSARY

adaptation an adjustment in natural or human systems to a new or changing environment. Adaptation to climate change refers to adjustment in natural or human systems in response to actual or expected climatic changes.

aerosols tiny bits of liquid or solid matter suspended in air. They come from natural sources such as erupting volcanoes and from waste gases emitted from automobiles, factories, and power plants. By reflecting sunlight, aerosols cool the climate and offset some of the warming caused by greenhouse gases.

albedo the relative reflectivity of a surface. A surface with high albedo reflects most of the light that shines on it and absorbs very little energy; a surface with a low albedo absorbs most of the light energy that shines on it and reflects very little.

anthropogenic emissions made by people or resulting from human activities. This term is usually used in the context of emissions that are produced as a result of human activities.

atmosphere the thin layer of gases that surround the Earth and allow living organisms to breathe. It reaches 400 miles (644 km) above the surface, but 80 percent is concentrated in the troposphere—the lower 7 miles (11 km) above the Earth's surface.

biodiversity different plant and animal species.

biofuel a fuel produced from organic matter or combustible oils produced by plants. Examples of biofuel include alcohol, black liquor from the paper-manufacturing process, wood, and soybean oil.

biomass the total mass of living organisms in a given area or volume; dead plant material can be included as dead biomass.

black carbon soot or charcoal and/or possible light-absorbing refractory organic matter.

bleaching (coral) the loss of algae from corals that causes the corals to turn white. This is one of the results of global warming and signifies a die-off of unhealthy coral.

cap and trade the cap and trade system involves the trading of emission allowances, where the total allowance is strictly limited or capped. Emissions trading is an administrative approach used to control pollution by providing economic incentives for achieving reductions in the emissions of pollutants. A company is allowed to have a specified level of pollution that they can sell and trade. If a company exceeds their limit, they can buy credits to decrease global warming elsewhere.

carbon a naturally abundant nonmetallic element that occurs in many inorganic and in all organic compounds.

carbon capture and storage (CCS) a process consisting of separation of CO_2 from industrial and energy-related sources, transport to a storage location, and long-term isolation from the atmosphere.

carbon cycle the term used to describe the flow of carbon (in various forms, such as carbon dioxide) through the atmosphere, ocean, terrestrial biosphere, and lithosphere.

carbon dioxide a colorless, odorless gas that passes out of the lungs during respiration. It is the primary greenhouse gas and causes the greatest amount of global warming.

carbon sink an area where large quantities of carbon are built up in the wood of trees, in calcium carbonate rocks, in animal species, in the ocean, or any other place where carbon is stored. These places act as a reservoir, keeping carbon out of the atmosphere.

chaos theory a theory to explain the nonlinear, deterministic behavior of certain systems. A dynamical system such as the climate system, governed by nonlinear deterministic equations, may exhibit erratic or chaotic behavior in that very small changes in the initial state of the system in time lead to large and apparently unpredictable changes in its temporal evolution. Chaotic behavior may limit a model's predictability.

climate the usual pattern of weather that is averaged over a long period of time.

Glossary

climate model   a quantitative way of representing the interactions of the atmosphere, oceans, land surface, and ice. Models can range from relatively simple to extremely complicated.

climate sensitivity   the equilibrium change in the annual mean global surface temperature following a doubling of the atmospheric equivalent carbon dioxide concentration.

climate system   The highly complex system consisting of five major components: the atmosphere, the hydrosphere, the cryosphere, the land surface, and the biosphere, and the interactions between them. The climate system evolves in time under the influence of its own internal dynamics and because of external forcings such as volcanic eruptions, solar variations, and anthropogenic forcings, such as the changing composition of the atmosphere and land use change.

climatologist   a scientist who studies the climate.

concentration   the amount of a component in a given area or volume. In global warming, it is a measurement of how much of a particular gas is in the atmosphere compared to all of the gases in the atmosphere.

condense   the process that changes a gas into a liquid.

convection   vertical motion driven by buoyancy forces arising from static instability, usually caused by near-surface cooling or increases in salinity in the case of the ocean and near-surface warming in the case of the atmosphere.

cryosphere   the component of the climate system consisting of all snow, ice, and frozen ground (including permafrost) on and beneath the surface of the Earth and oceans.

deforestation   the large-scale cutting of trees from a forested area, often leaving large areas bare and susceptible to erosion.

ecological   the protection of the air, water, and other natural resources from pollution or its effects. It is the practice of good environmentalism.

ecosystem   a community of interacting organisms and their physical environment.

emissions   the release of a substance (usually a gas when referring to the subject of climate change) into the atmosphere.

energy balance   the difference between the total incoming and total outgoing energy. If this balance is positive, warming occurs; if it is negative, cooling occurs.

evaporation   the process by which a liquid, such as water, is changed to a gas.

feedback   a change caused by a process that, in turn, may influence that process. Some changes caused by global warming may hasten the process of warming (positive feedback); some may slow warming (negative feedback).

feedstock   raw material for a processing or a manufacturing industry.

forcings   mechanisms that disrupt the global energy balance between incoming energy from the Sun and outgoing heat from the Earth. By altering the global energy balance, such mechanisms force the climate to change. Today, anthropogenic greenhouse gases added to the atmosphere are forcing climate to behave as it is.

fossil fuel   an energy source made from coal, oil, or natural gas. The burning of fossil fuels is one of the chief causes of global warming.

glacier   a mass of ice formed by the buildup of snow over hundreds and thousands of years.

global warming   an increase in the temperature of the Earth's atmosphere, caused by the buildup of greenhouse gases. This is also referred to as the enhanced greenhouse effect caused by humans.

global warming potential (GWP)   a measure of how much a given mass of greenhouse gas is estimated to contribute to global warming. It is a relative scale, which compares the gas in question to that of the same mass of carbon dioxide, whose GWP is equal to 1.

greenhouse effect   the natural trapping of heat energy by gases present in the atmosphere, such as carbon dioxide, methane, and water vapor. The trapped heat is then emitted as heat back to the Earth.

greenhouse gas   a gas that traps heat in the atmosphere and keeps the Earth warm enough to allow life to exist.

gross domestic product   the monetary value of all goods and services produced within a nation.

hybrid car   any vehicle that employs two sources of propulsion, especially a vehicle that combines an internal combustion engine with an electric motor.

IPCC (Intergovernmental Panel on Climate Change)   This is an organization consisting of 2,500 scientists that assesses information in the scientific and technical literature related to the issue of climate change. The United Nations Environment Programme and the World Meteorological Organization established the IPCC jointly in 1988.

land use   the management practice of a certain land cover type. Land use may be such things as forest, arable land, grassland, urban land, and wilderness.

methane   a colorless, odorless, flammable gas that is the major ingredient of natural gas. Methane is produced wherever decay occurs and little or no oxygen is present.

mitigation   technological change and substitution that reduces inputs and emissions per unit of output. Although several social, economic, and technological policies would produce an emission reduction, with respect to climate change, mitigation means implementing policies to reduce GHG emissions and enhance sinks.

model   a working hypothesis or precise simulation, by means of description, statistical data, or analogy, of a phenomenon or process that cannot be observed directly or that is difficult to observe directly. Models may be derived by various methods, such as by computer.

nitrogen   as a gas, nitrogen takes up 80 percent of the volume of the Earth's atmosphere. It is also an element in substances such as fertilizer.

nitrous oxide   a heat-absorbing gas in the Earth's atmosphere. Nitrous oxide is emitted from nitrogen-based fertilizers.

ozone   a molecule that consists of three oxygen atoms. Ozone is present in small amounts in the Earth's atmosphere at 14 to 19 miles (23–31km) above the Earth's surface. A layer of ozone makes life possible by shielding the Earth's surface from most harmful ultraviolet rays. In the lower atmosphere, ozone emitted from auto exhausts and factories is an air pollutant.

parts per million (ppm)   the number of parts of a chemical found in one million parts of a particular gas, liquid, or solid.

peat   peat is formed from dead plants, typically *Sphagnum* mosses, which are only partially decomposed due to the permanent submergence in water and the presence of conserving substances such as humic acids.

permafrost   permanently frozen ground in the Arctic. As global warming increases, this ground is melting.

photosynthesis   the process by which plants make food using light energy, carbon dioxide, and water.

power grid   the entire system upon which power travels from the power generation plant to its final destination. Also referred to as a power distribution grid, power is generated at the power plant and travels through transmission substations and lines, and distribution grids until it reaches homes, businesses, and other final destinations where it is used as electricity to operate lights, furnaces, and many other applications.

protocol   the terms of a treaty that have been agreed to and signed by all parties.

proxy   a proxy climate indicator is a local record that is interpreted, using physical and biophysical principles, to represent some combination of climate-related variations back in time. Climate-related data derived in this way are referred to as proxy data. Examples of proxies include pollen analysis, tree ring records, characteristics of corals, and various data derived from ice cores.

radiation   the particles or waves of energy.

remote sensing   the collection of information about an object by a recording device that is not in physical contact with it. It is the collection of reflected or radiated electromagnetic energy from the Earth and uses cameras, infrared detectors, microwave frequency receivers, and radar systems. It can be collected from both aircraft and satellite platforms.

Glossary

renewable something that can be replaced or regrown, such as trees, or a source of energy that never runs out, such as solar energy, wind energy, or geothermal energy.

resources the raw materials from the Earth that are used by humans to make useful things.

satellite any small object that orbits a larger one. Artificial satellites carry instruments for scientific study and communication. Imagery taken from satellites is used to monitor aspects of global warming such as glacier retreat, ice cap melting, desertification, erosion, hurricane damage, and flooding. Sea-surface temperatures and measurements are also obtained from man-made satellites in orbit around the Earth.

sequestration carbon storage in terrestrial or marine reservoirs. Biological sequestration includes direct removal of CO_2 from the atmosphere through land-use change, afforestation, reforestation, carbon storage in landfills, and practices that enhance soil carbon in agriculture.

simulation a computer model of a process that is based on actual facts. The model attempts to mimic, or replicate, actual physical processes.

sinks any process, activity, or mechanism that removes a greenhouse gas or aerosol or a precursor of a greenhouse gas or aerosol from the atmosphere.

spatial resolution (model) the level of detail a model has, referring to how far apart the x/y points in the grid are spaced. The closer the spacing, the more data in the model, making it more detailed and discerning.

sustainable development the concept of sustainable development was introduced in the World Conservation Strategy (UICN 1980) and had its roots in the concept of a sustainable society and in the management of renewable resources.

temperate an area that has a mild climate and different seasons.

thermal something that relates to heat.

tropical a region that is hot and often wet (humid). These areas are located around the Earth's equator.

troposphere the bottom layer of the atmosphere, rising from sea level up to an average of about 7.5 miles (12 km).

weather the conditions of the atmosphere at a particular time and place. Weather includes such measurements as temperature, precipitation, air pressure, and wind speed and direction.

FURTHER RESOURCES

BOOKS

Christianson, Gale. *Greenhouse: The 200-Year Story of Global Warming.* New York: Walker, 1999. Looks at the enhanced greenhouse effect worldwide after the industrial revolution and outlines the consequences to the environment.

Dow, Kirstin, and Thomas E. Downing. *The Atlas of Climate Change: Mapping the World's Greatest Challenge.* Los Angeles: University of California Press, 2006. This publication offers maps and geographic statistics and information on climate change, global warming, economics, and other related scientific topics worldwide.

Friedman, Katherine. *What If the Polar Ice Caps Melted?* Danbury, Conn.: Children's Press, 2002. Focuses on environmental problems related to the Earth's atmosphere, including global warming, changing weather patterns, and their effects on ecosystems.

Gelbspan, Ross. *The Heat Is On: The High Stakes Battle over Earth's Threatened Climate.* Reading, Mass.: Addison Wesley, 1997. This work offers a look at the controversy environmentalists often face when they deal with fossil fuel companies.

Harrison, Patrick "GB," Gail "Bunny" McLeod, and Patrick G. Harrison. *Who Says Kids Can't Fight Global Warming.* Chattanooga, Tenn.: Pat's Top Products, 2007. Offers real solutions that everybody can do to help solve the world's biggest air pollution problems.

Houghton, John. *Global Warming: The Complete Briefing.* New York: Cambridge University Press, 2004. This book outlines the scientific basis of global warming and describes the impacts that climate change will have on society. It also looks at solutions to the problem.

Langholz, Jeffrey. *You Can Prevent Global Warming (and Save Money!): 51 Easy Ways.* Riverside, N.J.: Andrews McMeel Publishing, 2003.

Aims at converting public concern over global warming into positive action to stop it by providing simple, everyday practices that can easily be done to minimize it, as well as save money.

McKibben, Bill. *Fight Global Warming Now: The Handbook for Taking Action in Your Community.* New York: Holt Paperbacks, 2007. Provides the facts of what must change to save the climate. It also shows how everyone can act proactively in their community to make a difference.

Pringle, Laurence. *Global Warming: The Threat of Earth's Changing Climate.* New York: SeaStar Publishing Company, 2001. Provides information on the carbon cycle, rising sea levels, El Niño, aerosols, smog, flooding, and other issues related to global warming.

Ruddiman, William F. *Earth's Climate: Past and Future.* New York: W. H. Freeman and Company, 2001. Takes a detailed look at the history of the Earth's climate and the forces that have shaped it over time.

Thornhill, Jan. *This Is My Planet—the Kids Guide to Global Warming.* Toronto, Ontario: Maple Tree Press, 2007. Offers students the tools they need to become ecologically oriented by taking a comprehensive look at climate change in polar, ocean, and land-based ecosystems.

Weart, Spencer R. *The Discovery of Global Warming (New Histories of Science, Technology, and Medicine).* Cambridge, Mass.: Harvard University Press, 2004. Traces the history of the global warming concept through a long process of incremental research rather than a dramatic revelation.

JOURNALS

American Wind Energy Association. "Wind Energy and Climate Change: A Proposal for a Strategic Initiative" (October 1997). Available online. URL: http://www.ecoiq.com/onlineresources/anthologies/energy/wind.html. Accessed March 20, 2009. Discusses cost-effective methods for supplying electricity to rural villages via renewable wind energy.

Further Resources

Broder, John M. "Democrats Unveil Climate Bill." *New York Times* (4/1/09). Available online. URL: http://www.nytimes.com/2009/04/01/us/politics/01energycnd.html?hp. Accessed January 23, 2009. Presents the viewpoint of global warming and politics.

———. "EPA Clears Way for Greenhouse Gas Rules." *New York Times* (4/18/09). Available online: URL: http://www.nytimes.com/2009/04/18/science/earth/18endanger.html. Accessed May 28, 2009. This presents the new ruling by the EPA in an effort to control greenhouse gases.

Choi, Charles Q. "The Energy Debates: Clean Coal." *Live*Science (12/5/08). Available online. URL: www.ivescience.com/environment/081205-energy-debates-clean-coal.html. Accessed February 22, 2009. Discusses whether or not the clean coal technology performs up to its expectations.

Dean, Cornelia. "The Problems in Modeling Nature, with Its Unruly Natural Tendencies." *New York Times* (2/20/07). Available online. URL: http://www.nytimes.com/2007/02/20/science/20book.html. Accessed May 2, 2009. Discusses the inherent limits of mathematical models and appropriate assumptions concerning their usage.

Flook, Simon. "China Set to Build 562 New Coal Plants—Kyoto in Perspective." *The Politic* (1/17/07). Available online. URL: www.the-politic.com/archives/2007/01/17/china-set-to-build-562-new-coal-plants/. Accessed January 16, 2009. Discusses air pollution concerns in China and the disastrous effect that will have on global warming if they do not use renewable energy sources, but rely on fossil fuels instead as they industrialize.

Gelling, Peter, and Andrew C. Revkin. "Climate Talks Take on Added Urgency after Report." *New York Times* (12/3/07). Accessed online. URL: www.nytimes.com/2007/12/03/world/asia/03bali.html?pagewanted=pring. Accessed January 23, 2008. Discusses the need to cut greenhouse emissions in preparation for the Bali conference, which will discuss what the global plan of action will be after the Kyoto Protocol expires in 2012.

Greinel, Hans. "Japan to Fight Global Warming by Pumping Carbon Dioxide Underground." *USA Today* (6/26/06). Available online. URL: http://content.usatoday.com/topics/more+stories/ Places,%20Geography/Countries/Norway/48. Accessed April 25, 2009. Explores the option of carbon sequestration as a viable way to counteract the effects of global warming.

Houghton, R. A., J. L. Hackler, and K. T. Lawrence. "The U.S. Carbon Budget: Contributions From Land-use Change." *Science* (7/23/99) 285, no. 5,427: 574–578. Discusses how the net carbon flux related to U.S. lands offsets 10 to 30 percent of the United States' fossil fuel emissions.

Kanter, James. "Europe May Ban Imports of Some Biofuel Crops." *New York Times* (1/15/08). Available online. URL: www.nytimes.com/2008/01/15/business/worldbusiness/15biofuel.html. Accessed March 24, 2009. Discusses which crops the European Union will not import and why they feel those crops add to the problem of global warming, not solve it.

Kaufman, Leslie. "Dissenter on Warming Expands His Campaign." *New York Times* (4/10/09). Available online. URL: www.nytimes.com/2009/04/10/us/politics/10morano.html?pagewanted=print. Accessed May 28, 2009. Discusses why it is important when reading about global warming to make sure the source is reliable and believable.

Kerr, Richard A. "Global Warming: Rising Global Temperature, Rising Uncertainty." *Science* (4/13/01) 192–194. Looks at climate modeling and the uncertainties currently associated with it. It also discusses important rules to understand when interpreting climate models.

Krauss, Clifford. "As Ethanol Takes Its First Steps, Congress Proposes a Giant Leap." *The New York Times* (12/18/07). Available online. URL: www.nytimes.com/2007/12/18/washington/18ethanol.html?pagewanted=print. Accessed January 23, 2008. Discusses the plans Congress has for renewable energy in order to reduce the nation's heavy reliance on foreign oil.

LaGesse, David. "The PC's Dirty Little Secret: It Wastes Power Shamelessly." *U.S. News and World Report* (4/17/08). Available online.

URL: http://www.usnews.com/articles/business/technology/2008/04/17/the-pcs-dirty-little-secret-it-wastes-power-shamelessly.html. Accessed May 9, 2009. This review discusses the real energy use of a PC and easy practices that can be followed to conserve energy and lower power bills.

———. "Small Moves You Can Take at Home to Conserve." *U.S. News and World Report* (4/17/08). Available online. URL: http://www.usnews.com/articles/business/technology/2008/04/17/small-moves-you-can-take-at-home-to-conserve.html. Accessed May 9, 2009. Discusses simple ways to save money on electricity, such as products like the Kill A Watt, the Energy Detective, the Bye Bye Standby, the Solatube, and the Voltaic Generator.

Lavelle, Marianne. "Conservation Can Mean Profits for Utilities." *U.S. News and World Report* (4/17/08). Available online. URL: http://www.usnews.com/articles/business/technology/2008/04/17/conservation-can-mean-profits-for-utilities.html. Accessed May 9, 2009. Discusses a new trend that encourages utility companies not to expand, but instead to urge customers to conserve energy.

———. "Three Ways Businesses Can Save on Power." *U.S. News and World Report* (4/17/08). Available online. URL: http://www.usnews.com/articles/business/technology/2008/04/17/three-ways-businesses-can-save-on-power.html. Accessed May 9, 2009. Discusses practical methods that factories and offices can use to become more energy efficient and save money.

———. "Green, Not Sacrifice, Is the Political Word." *U.S. News and World Report* (4/17/08). Available online. URL: http://www.usnews.com/articles/business/technology/2008/04/17/green-not-sacrifice-is-the-political-word.html. Accessed May 9, 2009. Discusses the perception today of the environmentalism movement focusing more on the positive aspects of being green instead of changes being viewed negatively as a personal sacrifice.

Mankiw, N. Gregory. "One Answer to Global Warming: A New Tax." *New York Times* (9/16/07). Available online. URL: www.

nytimes.com/2007/09/16/business/16view.html?pagewanted=print. Accessed April 22, 2009. Provides information on business options to impose a new tax to combat global warming that would fairly tax citizens, leaving the distribution of total tax burden basically unchanged.

New York Times. The "Winners and Losers in a Changing Climate." (4/2/07). Available online. URL: http://www.nytimes.com/2007/04/02/us/20070402_CLIMATE_GRAPHIC.html. Accessed May 13, 2009. Discusses which countries will be hit the hardest with the negative effects of global warming and will have the most adjusting to do. It also gives examples of what some countries are already doing to mitigate the future effects of global warming.

———. "The Scientists Speak." (11/20/07). Available online. URL: http://www.nytimes.com/2007/11/20/opinion/20tue1.html. Accessed May 25, 2009. This editorial discusses the latest scientific evidence and why it needs to be used by Washington to set forth climate policy that deals effectively with the issue of global warming.

———. "The One Environmental Issue." (1/1/08). Available online. URL: www.nytimes.com/2008/01/01/opinion/01fuel.html?pagewanted=print. Accessed January 23, 2008. Gives an overview of how the global warming issue was dealt with politically in the past and how it is being looked at today and why.

Reuters (London). "Multinationals Fight Climate Change." *New York Times* (1/21/08). Available online. URL: http://www.nytimes.com/2008/01/21/business/21green.html. Accessed 4/26/09. Looks at the joint efforts of 11 companies using renewable energy.

Revkin, Andrew C. "A New Middle Stance Emerges in Debate over Climate." *New York Times* (1/1/07). Available online. URL: http://www.nytimes.com/2007/01/01/science/01climate.html?_r=2&oref=slogin. Accessed May 25, 2009. Presents the opinion of a new group on the state of global warming that is neither far left or right, but the middle ground.

Further Resources

———. "Connecting the Global Warming Dots." *New York Times* (1/14/07). Available online. URL: http://www.nytimes.com/2007/01/14/weekinreview/14basics.html. Accessed January 15, 2009. Discusses the anthropogenic contributions to global warming.

Rosenthal, Elisabeth, and Andrew C. Revkin. "Science Panel Calls Global Warming 'Unequivocal.'" *New York Times* (2/3/07). Available online. URL: http://www.nytimes.com/2007/02/03/science/earth/03climate.html. Accessed May 25, 2009. Compares the force and certainty the IPCC's fourth report is written with to their earlier reports and the accelerated seriousness of the global warming issue.

Sachs, Jeffrey D. "Keys to Climate Protection." *Scientific American* (April 2008). Available online. URL: http://www.scientificamerican.com/article.cfm?id=technological-keys-to-climate-protection-extended. Accessed May 28, 2009. Discusses why the creation and implementation of new technology is critical to fight global warming with before it is too late.

Stolberg, Sheryl Gay. "Bush Sets Greenhouse Gas Emissions Goal." *New York Times* (4/17/08). Available online. URL: http://www.nytimes.com/2008/04/17/washington/17bush.html. Accessed May 25, 2009. Discusses the action that needs to be taken to avoid the worst of the greenhouse effect and global warming.

Thompson, Andrea, and Ker Than. "Timeline: The Frightening Future of Earth." *Live*Science. (4/19/07). Available online. URL: www.livescience.com/environment/070419_earth_timeline.html. Accessed May 13, 2009. Presents future predictions as to what the Earth's environment will be like from now until the 22nd century.

Wald, Matthew L. "New Ways to Store Solar Energy for Nighttime and Cloudy Days." *New York Times* (4/15/08). Available online. URL: http://www.nytimes.com/2008/04/15/science/earth/15sola.html. Accessed March 22, 2009. Discusses innovative ways to capture the Sun's heat via solar thermal systems.

Weber, Elke U. "Experienced-based and Description-based Perceptions of Long-term Risk: Why Global Warming Does Not Scare Us (Yet)." *Climatic Change* (2006) 77: 103–120. Discusses how people in general look at risk and how they prioritize decisions based on that perceived risk.

INDEX

Italic page numbers indicate illustrations or maps. Page numbers followed by *c* denote chronology entries; page numbers followed by *t* denote tables, charts, or graphs.

A

accuracy, balance v. 98–99
ACES. *See* American Clean Energy and Security Act of 2009
action, taking 183–184. *See also* solutions
active solar systems 133
adaptation 23, 28–30, 198–199
adaptive management 177
Advisory Group on Greenhouse Gases (AGGG) 76
aerosols 173
agriculture 29, 89, 112–113
Ahrenkilde-Hansen, Pia 18
airlines 200*t*
air pollution 123, 126–127
albedo 73, 165
Alley, Richard B. 118
allowance-based markets (cap and trade) 59–60
American Clean Energy and Security Act of 2009 (ACES) 35, 52–53
Angel, Roger 195–196
Annex I countries 7–9, 12, 13, 57, 58
Annex II countries 8
Antarctic Ice Sheet 213, 234*c*, 236*c*
anthropogenic emissions 2–4, 25–28, 31, 179
appliances, household 191, 201*t*–202*t*
AR4. *See* Fourth Assessment Report of the IPCC
Arctic ecosystem 73–75
Árnason, Bragi 84, 86
Arrhenius, Svante 233*c*
Arsonval, Jacques-Arsène d' 154
assigned amount units (AAUs) 58
Atlantic Ocean conveyor belt 27. *See also* thermohaline circulation
atmosphere general circulation models (AGCMs) 170*t*
Australia 18, 43, 89
automobiles 80–81, 84–86, 186–190
automotive maintenance 200*t*
aviation emissions 80

B

Baker, James 98
balance, journalistic 97–100
Ban Ki-Moon 90
Barroso, José Manuel 82
Bebb, Adrian 151
Becquerel, Edmund 134
beef 209*t*
believers in global warming 116–121, 120*t*–121*t*
Benn, Hilary 17
binary power plant 141
biodiesel 149
biodiversity loss 29
biomass 46–48, 148–152, 157–158, *158*
biopower 148–152
birds 145
black carbon 157
Black Sunday 233*c*
Boston, Massachusetts 5
Boxer, Barbara 48
Brady, Aaron 47
Brown, Gordon 83
Bureau of Land Management, U.S. (BLM) 125–126
Bush, George H. W., and administration 8
Bush, George W., and administration
 Consolidated Appropriations Act of 2008 49
 Kyoto Protocol 9–10, 12, 15–18
 renewable energy programs 124–125
 silencing of critics 96
 skepticism on climate change 237*c*

C

California 62, 64–66, 144
California Air Resources Board (CARB) 65
California Climate Bill 65
Callendar, Guy S. 234*c*

255

Canada 5, 88
Canada Bay, Australia 89
Canadian Wildlife Service (CWS) 73
cap and trade 55–62
   and ACES 53
   allowance-based markets 59–60
   carbon credits 60*t*
   economics of 61–62
   Kyoto Protocol 12–13
   Lieberman-McCain proposal 237*c*
   project-based markets 57–59
car(s). *See* automobiles
Car Allowance Rebate System (CARS) 186–188, 188*t*–189*t*
carbon capture and storage (CCS)
   ACES funding 36
   artificial tree proposal 197
   DOE research 155–156
   Iceland 86
   Japan 87–88
   Norway 86–87
   U.S. research needs 192
carbon credits 12–13, 59–60, 60*t*, 191–192
carbon dioxide (CO_2)
   from deforestation 191
   ice core studies 235*c*
   and IPCC Working Group reports 26, 27, 31
   Kyoto Protocol reduction commitments 8
   levels (1800–1870) 233*c*
   levels (1950–1970) 234*c*
   levels (2003) 237*c*
   from peat destruction 151
   persistence in atmosphere 186
   and skeptics' argument against global warming 112
   and temperature 214*t*
   and unknowns in climate modeling 180
Carbon Disclosure Project 184
carbon flux 112
Carbon Sciences 157
carbon sinks 112
carbon storage 212
carbon tax 70
Carter, Rob 112–113
Cash for Clunkers (Car Allowance Rebate System) 186–188, 188*t*–189*t*
CCS. *See* carbon capture and storage
Central Europe floods (2002) 236*c*
Chandler, Mark 171
chaos theory 174

Charlson, Robert 178
Chevy Volt 189
Chicago, Illinois 5
China 12, 15, 18, 218–219, 237*c*
chlorofluorocarbons (CFCs) 77
Chukchi Sea 217
cirrus clouds 175–176
Clapp, Philip 17
Claussen, Eileen 86
Clean Air Act 45, 46, 48, 50
Clean and Diversified Energy Initiative (CDEi) 67
clean development mechanism (CDM) 13, 57–58
Clear Skies and Global Climate Change Initiatives 15–16
Climate Action Network of Australia 43
Climate Change Action Plan (London, England) 90
climate investment funds (CIFs) 20
climate modeling 159–181, *166*
   aerosols and 173
   clouds and 173–176
   Community Climate System Model 220–222
   confidence and validation 171–172
   educational programs 171
   error amplification 177–178
   fundamentals of 162–172, *166*, 170*t*
   and GHGs 4
   history of 160–162, 163*t*
   Met Office Hadley Centre configurations 170*t*
   midlevel warming problem 107–108
   modeling the response to change 165–167
   nature's inherent unruliness 176–177
   "petascale" computer modeling 220–221
   physics of 164
   simplifying the climate system 164–165
   solar variability 172–173
   testing of models 169
   uncertainties and challenges 172–181
   unknowns 180–181
Climatic Research Unit (University of East Anglia) 105
climatologists 160
Clinton, Bill, and administration 9, 14, 18
clotheslines 203*t*
clothing 208*t*
clouds 161, 162, 173–176, 196
CO_2. *See* carbon dioxide

coefficients 176
Community Climate System Model (CCSM) 220–222
community education 206–207
computer modeling 160–161, 163, 219–220. *See also* climate modeling
computers, energy saving with 206*t*
Conference of the Parties (of UNFCCC) 9
confidence, degrees of 25
Congress, U.S. 8, 46, 48–49, 192. *See also* Senate, U.S.
Connolley, William 100–101
Consolidated Appropriations Act of 2008 49–51, 51*t*
controversy over global warming. *See* debate over global warming
Copenhagen Conference on Climate Change (2009) 70, 238*c*
corn 47–48
Corzine, Jon 83
cryogenic coolers 155
cultural values 92–95

D

dams 146–147, 153
Davies, Kert 99
Davis, Gray 65
debate over global warming 103–121
    believers on far left 116–121
    middle ground 115–116
    modern climate consensus 104–109
    skeptics on far right 109–115
deforestation 191, 234*c*
degrees of confidence 25
"Democrats Unveil Climate Bill" (*New York Times* article) 35
Denmark 144
developing countries 12, 15, 217
direct-fired biopower 148–149
DOE. *See* Energy, U.S. Department of
Doniger, David 17
drought 26, 28, 30, 234*c*, 236*c*–237*c*
Duke Energy Corporation 45
dust bowl 233*c*
dynamic processes 165

E

Earth Summit 1992 (Rio de Janeiro) 8, 235*c*
Earth Summit 2002 (Johannesburg) 78
EAUs (EU allowance units) 13
ecological footprint 2, 182

economic issues
    benefits of green energy 125–128
    and Bush administration's alternatives to Kyoto 16
    cap and trade 61–62
    mitigation projects 68–71
education 171, 206–207
Educational Global Climate Model (EdGCM) 171
election of 2000 32–33
election of 2008 34–35
electric cars 189
electricity generation/distribution
    geothermal 141
    modernization 190
    pyrolysis oil 150
    solar power 134–136
    wind energy 142–146, *143*
emergent qualities 164
emigration 44
emission reduction units (ERUs) 58
emissions trading schemes. *See* European Union Greenhouse Gas Emissions Trading System
endangered species 148, 151
Energy, U.S. Department of (DOE) 67, 88, 126, 136, 192
energy balance models (EBMs) 167
energy budget 174, 175
energy efficiency 190–191, 201*t*–204*t*
ENERGY STAR® program 191, 238*c*
energy use 123
enhanced greenhouse effect 234*c*
Entman, Robert 98
Environmental Action Plan (California) 62, 64
Environmental Defense Fund (EDF) 57
Environmental Protection Agency, U.S. 44–46, 49–51, 126
Essential Requirements for Mandatory Reporting (ERMR) 66
ethanol 46–48, 149, 157
EU ETS. *See* European Union Emissions Trading System
Europe 106–107, 151–152, 237*c*
European Climate Change Programme (ECCP) 79–82
European Commission 78–81
European Union (EU) 151, 211
European Union Greenhouse Gas Emissions Trading System (EU ETS) 13, 59, 79–80
experiments, climate modeling 160

extinctions 29. *See also* endangered species
extreme weather events 106–107

F

Fairness and Accuracy in Reporting (FAIR) 97
feedback 160, 174
Fein, Jay 221
finite pool of worry hypothesis 94
First Assessment Report (FAR) 22
Flannery, Mark 46
flash tank 141
flooding 29–30, 107, 218, 236c
food production 28, 47–48, 209t. *See also* agriculture
food shortages 28–29, 42
forcings 23, 163
forests 191–192
fossil fuels 151, 237c
Fourier, Jean-Baptiste-Joseph 103, 233c
Fourth Assessment Report of the IPCC (AR4) 3, 25–31, 90, 117–118, 178–179, 237c–238c
Framework Convention on Climate Change 235c, 236c
free market 16
freshwater 40–43, 181
fuel cell vehicles 84–86
fuel efficiency 53–54, 80–81, 186, 187
future issues 210–222
futures market 60

G

G-8 countries 19–20, 20t, 237c
gasification 149, 157–158
gasoline price shock (2008) 94
general circulation models (GCMs) 167–172, 170t
General Motors 189
geochemistry 23
geoengineering 193–198, *194*
geological carbon sequestration. *See* carbon capture and storage
Geostationary Meteorological Satellite-5 (GMS-5) 175
geothermal energy 85, 139–142, *140*
Germany 17, 144
GHGs. *See* greenhouse gases
Gislason, Sigurdur Reynir 86
glacial melt 5, 26, 28, 218
Global Change Research Program, U.S. 216

global climate computer model (GCCM) 171
global cooling 100–101
global dimming, artificially induced 193–194
Global Effects of Environmental Pollution Symposium (Dallas, 1968) 75
Global Warming Pollution Reduction Act of 2007 48–49
global warming potential (GWP) 118
Global Warming Wildlife Survival Act 49
Goddard Institute for Space Studies (GISS) 104, 105, 171, 213–214
Gore, Al
  election of 2001 32–33
  An Inconvenient Truth 33, 237c
  Kyoto Protocol 14
  National Journal attack on 111
  Nobel Peace Prize (2007) 185, 238c
  on renewable energy 69–70
  Sanders-Boxer bill 48
Grandia, Kevin 100
Great Britain 17
Great Plains Synfuels Plant (Beulah, North Dakota) 88
"green carbon" 156–157
green energy
  ACES 35
  biomass 148–152
  California legislation 65–66
  congressional funding priorities 192
  electricity production 190
  environmental benefits 122–130
  geothermal 139–142, *140*
  Gore proposals 69–70
  hydropower 146–148, *147*
  Iceland 84
  ocean energy 152–155
  solar energy 130–139, *131, 132, 135*
  wind energy 142–146, *143*
greenhouse effect 103, 175, 233c–235c
greenhouse gases (GHGs). *See also specific gases, e.g.:* carbon dioxide
  and ACES 52–53
  and Consolidated Appropriations Act 49–50, 51t
  current legislation 44
  emissions by country 14t
  emissions by sector 63t
  from energy production 123
  EU reduction recommendations 211
  and general circulation models 168
  government inaction on 33

Iceland's per capita emissions 84
and IPCC Working Group reports 24–25, 31
and Kyoto Protocol 8–10
legislative proposals 35
and nighttime temperatures 4
non-CO_2 157
per capita responsibility, worldwide *218*
and Sanders-Boxer bill 48–49
and UNFCCC 6–8
Greening Earth Society 110, 112
Greenland ice core studies 235*c*, 236*c*
Greenland ice sheet 213
Gregoire, Chris 66
grocery bags 209*t*
gross domestic product (GDP) 31, 61, 187
groundwater 140
Group of 8 (G-8). *See* G-8 countries
Gulf Stream 41–42
Gutzler, David 179

H

Hansen, James E. 72–73, 104
health costs of air pollution 126, 127
heat waves 5, 27, 30, 106–107, 237*c*
Hegerl, Gabriele 3
Heiligendamm Summit (2007) 19–20
Hill, Antonio 70–71
hindcasting 169
Holdren, John P. 118
home, actions to mitigate global warming *201*, 201–204
home energy audits 191
home heating, geothermal for 140–141
Houghton, R. A. 112
Howard, John 18
Huber, Peter 112
Hulme, Mike 116
human activity. *See* anthropogenic emissions
human brain 93
Hurricane Katrina 237*c*
hurricanes 26
hydrofluorocarbons (HFCs) 9
hydrogen fuel cell vehicles 84–86
hydropower 65–66, 146–148, *147*
hydrothermal reservoirs 139–140

I

ice age 233*c*
ice core studies 42, 234*c*–236*c*
ice flow 27, 212, 213
Iceland 84–86, *85*

Impacts, Adaptation and Vulnerability (WGII report) 23–24
Inconvenient Truth, An (film) 33, 237*c*
India 12, 15
industrialized nations 217
inertia 210, 212
infrared iris 175–176
input error 177–178
insurance industry 68–69
Intergovernmental Panel on Climate Change. *See* IPCC
International Carbon Action Partnership (ICAP) 82–83
International Council for Local Environmental Initiatives (ICLEI) 88–89
International Energy Agency (IEA) 20, 22
international issues
    cooperation 75–78
    international organizations' role 78–84
    Kyoto Protocol reactions 17–19
    politics 72–91
    progress of individual countries 84–91
    sustainability efforts 88–91
International Monetary Fund (IMF) 68
International Partnership for Energy Efficiency Cooperation 20
IPCC (Intergovernmental Panel on Climate Change) 20–31
    on CCS 88
    creation of 77, 235*c*
    Fourth Assessment Report 3, 25–31, 90, 117–118, 178–179, 237*c*–238*c*
    on global warming in U.S. and Canada 5–6
    and media 97–98
    mitigation cost estimates 69
    Nobel Peace Prize 33, 185, 238*c*
    skeptics' attack on reports 111
    temperature projections 212
    timetable for GHG emission reductions 90
    unequivocal nature of Fourth Report 117–118
    warming/sea-level rise projections 213
    Working Groups 22–31, *23, 24*
Iris hypothesis 175–176
iron hypothesis 195

J

Jackson, Lisa P. 46
Japan 87–89, 107

jobs creation 144–145, 152
Johannesburg Summit (2002) 78
joint implementation (JI) 13, 57, 58
journalistic balance 97–100

K

Karl, Thomas 105, 113
Kårstø, Norway 87
Katrina, Hurricane 237c
Keeling, Charles David 234c
Keeling Curve 75, 234c
Keller, Martin 47
Keohane, Nat 61–62
Kerry, John 33
Kiehl, Jeffrey 179
Kirtman, Ben 220, 222
Kristinsdóttir, Ásdis 84–85
Kroft, Steve 37
Kyoto Protocol 8–19, 9
   adoption of 237c
   Australian response 89
   cap and trade 56
   clean development mechanism 13, 57–58
   countries' positions on 11
   and European Climate Change Programme 79, 80
   inadequacy of 157
   international reaction 17–19
   joint implementation 13, 57, 58
   target CO_2 levels 69
   and UNFCCC 6, 236c
   U.S. response 14–17

L

Lackner, Klaus 196–197
LaHood, Ray 187, 188
landscaping 204t–205t
La Niña 106
Latham, John 196
Launder, Brian 196–198
Laurance, William 152
left-wing political viewpoint 116–121
legislation 44–54. *See also specific acts and laws*
Le Treut, Hervé 119
Levene, Lord Peter 69
Levitus, Sydney 119
Lieberman, Joseph 35, 237c
Lindzen, Richard 175, 176
Little Ice Age 233c
Livingstone, Ken 90

local food 209t
London, England 90
Los Angeles, California 5
Lubich, Chiara 222
Lunn, Nick 73–75

M

MacCracken, Michael 217
Mahlman, Jerry D. 179
Malawi 89
Manabe, Syukuro 161
Mann, Michael 178–179
market economy 16
Markey, Edward J. 35, 52
Marshall, Andrew 40
Martin, John 195
mathematics 160, 176–177
Mauna Loa CO_2 monitoring station (Hawaii) 234c
McCain, John 34–35, 237c
McCurdy, Dave 45
McKinsey Global Institute 69
meat 209t
media 92–102
   and changing scientific viewpoints 100–102
   human psychology/cultural values 92–95
   journalistic balance 97–100
   power of 95–97, 97
Merkel, Angela 17
Metcalf, Gilbert 70
methane (CH_4) 8, 26, 149
methanol 149
Met Office Hadley Center 107, 170t
mirrors, space-based 195–196
mitigation 55–71
   and AR4 30–31
   cap and trade 55–62
   economics of 68–71
   IPCC Working Group reports 23–25, 30–31
   state projects 62, 64–68
Mitigation of Climate Change (WGIII report) 30–31
modeling. *See* climate modeling
modular biomass systems 149
Moffic, H. Steven 92–93
Mongstad, Norway 87
Montreal Conference (2005) 13
Montreal Protocol of the Vienna Convention 77

Index

Morano, Marc 99–100
Munich Re (reinsurance) 69
Murphy, Terry 137, 138

N

National Aeronautics and Space Administration (NASA) 106, 162
National Assessment 216, 236c
National Climatic Data Center 105
National Greenhouse Gas Inventories Programme (IPCC-NGGIP) 6, 22, 25
National Journal 111
National Oceanic and Atmospheric Administration (NOAA) 105, 119
National Renewable Energy Laboratory (NREL) 157
national security 40–44
negative feedback 176
net metering 145
Newsom, Gavin 62, 64
New York City 5, 6
Nickles, Greg 19
Nishio, Masahiro 88
nitrous oxide (NO_x) 9
Nobel Peace Prize (2007) 33, 185, 238c
North, Gerald 178
Norway 86–87

O

Obama, Barack 36
    ACES 35
    Cash for Clunkers 186
    commitment to climate change action 238c
    fuel efficiency policy 53–54
    global warming outlook 36–40
ocean(s) 119, 195
ocean energy 152–155
ocean general circulation models (OGCMs) 170t
oceanic conveyor belt. *See* thermohaline circulation
ocean thermal energy conversion (OTEC) 155
O'Donnell, John S. 137
oil 233c, 238c
oil crisis (1973) 19, 127
"One Environmental Issue, The" *(New York Times* editorial) 32–33
Organization for Economic Co-operation and Development (OECD) 8, 22
Outreach Five (O5) 19

ozone (ground-level) 30
ozone (stratospheric) 76–77

P

Pachauri, Rajendra *21,* 22, 90, 117
paleoclimatology 22
palm oil 151
paper 208t
parametrization 165
passive solar energy 132–133
payroll tax 70
peat 151
perfluorocarbons (PFCs) 9
personal choices/actions 2–3, 208–209
"petascale" computer modeling 220–221
Peterson, Thomas 119
phenol 150
photosynthesis 148
photovoltaic cells 134–136, 138
physical science 25–27
Physical Science Basis (WGI report) 25–27
physics 160, 164
phytoplankton 195
Pickett, John 151–152
Pielke, Roger A., Jr. 116
Pilkey, Orrin H. 176–177
Pilkey-Jarvis, Linda 176
Pinatubo, Mount 105, 169
polar bears 73–75, *74*
politics of climate change in U.S. 32–54. *See also* Congress, U.S.
pollutants, climate modeling and 173
pollution permits 35
population 44
Portland, Oregon 6
Poulsen, Erik 66
poverty 217
power tower 137–138
precipitated calcium carbonate (PCC) 156–157
precipitation 27, 107, *181*
project-based markets (cap and trade) 57–59
proxies (proxy data) 107, 161
psychology, human 92–95
public awareness 206–207
Public Utility Regulatory Policies Act (PURPA) 129
pyrolysis oil 150, 158

R

radiation, clouds and 162

radiative-convective models (RCMs) 167–168
radiative processes 165
Randall, Doug 40, 42–43
Rayner, Steve 100
recession (2008–present) 70
recycling 183, 209*t*
regional climate models (RCMs) 170*t*
regulation 125–126
renewable energy. *See* green energy
Renewable Energy and Energy Efficiency Partnership (REEEP) 78–79
renewable energy and energy efficiency systems (REES) 78
research and development 219–220
    and Bush administration's alternatives to Kyoto 16
research on global warming 155–158, *158*
reservoirs 147–148
resolution. *See* spatial resolution
resources 44, 182
resource wars 42
right wing political viewpoint 109–115
Rio Summit (1992). *See* Earth Summit 1992
risk, human response to 93–95
Roberts, John G., Jr. 45
Robock, Alan 193–194

S

Sachs, Jeffrey D. 91, 219
Salter, Stephen 196
Sanders, Bernie 48
satellites 107–108, 162, 175
Schmidt, Gavin A. 162–163, 181
Schneider, R. Stephen 101
Schwartz, Peter 40, 42–43
Schwarzenegger, Arnold 37, 38, 62, 64
Science 112
sea ice loss 26, 27, 74–75
sea-level rise
    2003 studies 237*c*
    global projections by 2099 212
    IPCC Working Group report 27
    Shishmaref, Alaska 217
    thermal expansion 119
Seattle, Washington 19
Sellers, William 161
Senate, U.S. 14–15, 33, 35
Senate Foreign Relations Committee 8
sequestration 86, 155–156. *See also* carbon capture and storage
Serreze, Mark 73

ships, wind-powered 196
Shire of Yarra Ranges, Australia 89
Shishmaref, Alaska 217
"shock and trance" cycle 37–38
single action bias 94–95
skeptics, global warming 109–115, 114*t*
Smithsonian Tropical Research Institute 152
smog 30
Snøhvit, Norway, sequestration projects 86, 87
solar electric generating stations (SEGS) 134
solar energy 130–139, *131*, *132*, *135*, 155
solar farms 155
solar radiation, clouds and 162
solar thermal concentrating systems 133–134
Solar Two project *135*
solar variability 172–173
solutions
    adaptation strategies 198–199
    American ingenuity and research 192–193
    community 206–207
    electricity system modernization 190
    energy efficiency 190–191
    forest protection 191–192
    fuel efficient cars 186–190
    geoengineering projects 193–198, *194*
    at home *201*, 201–204
    impractical 193–198, *194*
    personal choices/actions 208–209
    practical 184, 186–193
    simple activities 199–209
    taking action 183–184
    transportation choices 200
    workplace choices 206
    in yard 204–205, *205*
SourceWatch 109, 112
Southeast Asia 151
space, mirrors in 195–196
spatial resolution 164, 165, *166*
Spitzer, Eliot 83
spring, arrival of 29
Standlee, Christopher G. 47
state mitigation projects 62, 64–68, 126
statistical-dynamical models (SDMs) 167, 168
Statoil-Hydro 87
steam turbine power plants 141
Steiner, Achim 91, 118
Stern, Sir Nicholas 68, 237*c*

Stern Review on the Economics of Climate Change 68
Stevens, John Paul 45
Stone, Peter 178
storage of solar energy 137–138, 150
storm surges 5, 6
stratospheric cooling 4, 113
Study of Critical Environmental Problems (SCEP) 75
Study of Man's Impact on Climate (SMIC) 75
sulfur 193–194
sulfur dioxide 193
sulfur hexafluoride (SF_6) 9
summer flows 5–6
sunspot cycle 173
supercomputers 220
Supreme Court, U.S. 44–45
surface processes 165
sustainable agriculture 150
sustainable development 23, 88–91
sustainable energy. *See* green energy
SUVs, fuel-efficient 186–190
Swiss Re (reinsurance) 69
Switzerland 218
syngas 158

T

target CO_2 levels 69
Task Force on National Greenhouse Gas Inventories (TFI) 25
tax credits 124, 138
technology 192–193, 219–222
temperature. *See also* global warming
  and CO_2 concentrations 214t
  global warming skeptics' use of discrepancies in data 113
  projected increase (2010–2029) 211
  unknowns in climate modeling 180
terrorism 40
testing of climate models 169
Thatcher, Margaret 77
thermal expansion of seawater 119
thermohaline circulation (THC) 27, 40–42, *41*, 180–181
tidal energy 153
tidal fences 153–154
tidal turbines 154
Toronto Conference 77
transportation choices to mitigate global warming 200
trees 204t, 205t
trees, imitation 196–198

Trenberth, Kevin 119
troposphere 113
Tyndall, John 233c

U

UCS. *See* Union of Concerned Scientists
UNCED (United Nations Conference on Environment and Development). *See* Earth Summit 1992
uncertainties, in climate modeling 172–181
UNFCCC. *See* United Nations Framework Convention on Climate Change
Union of Concerned Scientists (UCS)
  on complexity of global warming challenge 108–109, 116
  on economic benefits of renewable energy technology 126, 127
  on forest management 192
  on importance of immediate action on global warming 184, 186
  power plant study 190
  prioritizing of adaptation strategies 198–199
United Kingdom 78
United Nations 7, 17
United Nations Climate Change Conference (Copenhagen, 2009) 70, 238c
United Nations Climate Change Secretariat 13
United Nations Conference on Environment and Development (UNCED). *See* Earth Summit 1992
United Nations Conference on Green Cities 64
United Nations Environment Programme (UNEP) 21, 76
United Nations Framework Convention on Climate Change (UNFCCC) 4, 6–8, 22, 57–58, 82
United Nations Framework Convention on Climate Change Conference of the Parties 9
United States
  ACES 35, 52–53
  biomass potential 150, 152
  Consolidated Appropriations Act 49–51, 51t
  fuel efficiency policy 53–54
  global warming impact 5–6
  Global Warming Pollution Reduction Act 48–49

green energy 123–130
home solar energy 133
Kyoto Protocol 9–10, 14–17
legislation 44–54
national security 40–44
Barack Obama's outlook on global warming 36–40
political arena 32–54
regional winners and losers from global warming 216–219
solar power *132,* 138
state mitigation projects 62, 64–68
wind energy 144–145
United States Climate Action Partnership 36
United States Climate Change Science Program 113
United States National Assessment on the Potential Consequences of Climate Variability and Change 216, 236*c*

V

validation of climate modeling 171–172
values, cultural 92–95
Vienna Convention for the Protection of the Ozone Layer 76–77
volcanic eruptions 105–106
volcano, imitation 193–194

W

Warm-Biz 89
Washington State mitigation projects 66–67
wastewater recycling 89
water, in computer modeling 161–162
water heaters, solar 133
water shortages 5–6, 26, 28, 42
water vapor 175, 234*c*
wave energy 154

Waxman, Henry A. 35, 36, 52
Weber, Elke U. 93–95
Weber, Frank 189
Weiner, Jonathan 76
Western Climate Initiative (WCI) 66
Western Governors' Association (WGA) 67–68
Western Regional Air Partnership (WRAP) 67
Western Renewable Energy Zones (WREZ) 67
Wilkins Ice Shelf 238*c*
wind energy 127–130, 142–146, *143,* 155, 238*c*
wind farms 127
wind turbines 142–143
winter recreation 5
wood 148, 208*t*
Working Group I (WGI) 22–23, *23,* 25–27
Working Group II (WGII) 23–24, 28–30
Working Group III (WGIII) *24,* 24–25, 30–31
workplace choices to mitigate global warming 206
World Bank 20
World Climate Research Programme 234*c*
World Conference on the Changing Atmosphere: Implications for Global Security (Toronto Conference) 77
World Meteorological Organization (WMO) 21, 77
World Summit on Sustainable Development (WSSD) 78
Wunsch, Carl 115

Y

yards, climate change mitigation in 204–205, *205*